International Aid and China's Environment

Rapid economic growth in the world's most populous nation is leading to widespread soil erosion, desertification, deforestation and the depletion of vital natural resources. The scale and severity of environmental problems in China now threaten the economic and social foundations of its modernization.

International Aid and China's Environment analyses the relationship between international and local responses to environmental pollution problems in China. The book challenges the prevailing wisdom that weak compliance is the only constraint upon effective environmental management in China. It makes two contributions. First, it constructs a conceptual framework for understanding the key dimensions of environmental capacity. This is broadly defined to encompass the financial, institutional, technological and social aspects of environmental management. Second, the book details the implementation of donor-funded environmental projects in both China's poorer and relatively developed regions. Drawing upon extensive fieldwork, it seeks to explain how, and under what conditions, international donors can strengthen China's environmental capacity, especially at the local level. It will be of interest to those studying Chinese politics, environmental studies and international relations.

Katherine Morton is a Fellow in the Department of International Relations, Research School of Pacific and Asian Studies at the ANU in Australia.

Routledge Studies on China in Transition
Series Editor: David S. G. Goodman

International Aid and China's Environment

Taming the Yellow Dragon

Katherine Morton

Routledge
Taylor & Francis Group

LONDON AND NEW YORK

First published 2005 by Routledge

Published 2017 by Routledge
4 Park Square, Milton Park, Abingdon, Oxon OX14 4RN
605 Third Avenue, New York, NY 10017

*Routledge is an imprint of the Taylor & Francis Group, an informa
business*

Typeset in Times New Roman by
Newgen Imaging Systems (P) Ltd, Chennai, India

British Library Cataloguing in Publication Data
A catalogue record for this book is available from the British Library

Library of Congress Cataloging in Publication Data
A catalog record for this book has been requested

ISBN 978-0-4156-4868-4 (pbk)

Contents

Illustrations

Preface

This is a book about the relationship between international and local responses to environmental problems in China. It seeks to determine the effectiveness of international environmental aid by exploring the linkages between different donor approaches and local capacity. To a certain extent, the study was motivated by the theoretical debate over the role of the market versus citizen participation in environmental management. I was interested in finding out how these ideas could translate into practice in the context of a developing country that was undergoing considerable economic and social change. However, my initial theoretical inquiry soon became overshadowed by practical considerations and, in particular, the problem of weak local capacity. I became convinced that, in the case of China, the promotion of market or participatory values via the processes of international aid was not enough to bring about an improvement in environmental management. Instead, it seemed to me that both international donors and the Chinese government needed to focus more of their attention upon developing environmental capacity at the local (provincial, municipal and country level).

When I began the research in 1997 my initial inquiries led me to seriously reconsider the task that I had set myself. At the time, the Chinese government was still ill at ease with placing too much emphasis upon environmental protection. It was seen by many government officials as an unaffordable luxury. Although China appeared to be taking the environment seriously at the international level through its participation in environmental negotiations, it remained deeply sceptical of the motives of international donors. In the eyes of the Chinese government, environmental assistance was double edged. It offered much needed financial and technical assistance but, at the same time, provided richer nations with political leverage over China's future economic development. Scepticism appeared to run even deeper at the local level where hard evidence confirmed that economic realities far outweighed environmental concerns.

But as the research progressed it soon became clear that important changes were underway. My focus upon environmental aid projects provided an entry ticket into the day-to-day workings of environmental management across diverse regions of China. This, in turn, gave me an opportunity to establish a dialogue with various government agencies in Beijing that were keen to obtain insights into local environmental practices.

Mapping out the processes of environmental aid implementation proved to be a complex task; donor motives and interests were often obscure and difficult to identify, and local government attitudes and capabilities were rarely detectable on the surface but required patient probing and continuous assessment. The picture that emerged was messy. Yet, what became visible within it was a pattern of subtle changes in environmental capacity that were the culmination of local innovations, policy reforms in Beijing and the efforts of a small number of dedicated staff working in donor institutions in Tokyo, Washington, New York and Nairobi. This book seeks to draw attention to these changes, to make them more visible so that future efforts to manage environmental problems in China can be grounded in a better understanding of what actually works in particular settings.

One likes to think of oneself as an independent researcher. Yet in my experience this is rarely the case in practice. It is sobering to think how utterly dependent I have been on other people during the course of my research and preparation of the manuscript. I benefited considerably from the guidance of Gregory Noble, Lorraine Elliot and Yongjin Zhang. I would also like to thank Jonathan Unger and David Goodman for their generous advice and support, and Andrew Watson and Richard Louis Edmonds for their helpful suggestions over revisions. I owe a great deal to my colleagues within the Department of International Relations at the Research School of Pacific and Asian Studies, Australian National University, for their encouragement and constructive advice. A special thank you to Chris Reus-Smit for his rigorous criticism and never ending enthusiastic support. I am very grateful to Mary-Louise Hickey and Robin Ward for their wonderful editorial support. Robin deserves a special thank you for producing the index. My thanks also to Wynne Russell and Shogo Suzuki for their helpful editorial suggestions. I am also obliged to Kay Dancey for helping to produce the maps and on-line drawings, and to Darren Boyd for his help with the photographs.

During my fieldwork in China, the Centre for Environmental Sciences at Peking University and the Institute of Human Ecology in Beijing provided critical support. I received special encouragement from Chinese scholars working on environmental economics including Zhang Shiqui, Ma Zhong and Hu Tao. I owe a debt of gratitude to a large number of people working within international aid agencies who were generous with their time. A special thank you to Michio Hashimoto, Hideaki Hoshina, Keiichi Tango, Edouard Motte, Songsu Choi, Stuart Whitehead, Ai-Chin Wee, Miao Hongjun and Chris Radford.

I am especially grateful to the many local environmental officials in China who often went out of their way to provide useful information and encouraged me to pursue my research often in difficult circumstances. In particular, I would like to thank Song Diantang, Wang Xueyan, Tang Guimei, Yu Di, Wang Li, Xu Guo, Ding Yongfu, Mi Hua, Chen Xueming, Zheng Huiying, Yang Yingfeng, Jiang Renjie and Sun Hujun. I accept sole responsibility for any failures of omission or interpretation.

Last but by no means least, I am forever grateful to my husband Mark for his unconditional support and confidence in my work, and to my two young sons,

Daniel and Luke, for their patience and forbearance. Without their help this book would not have seen the light of day. I would like to dedicate the book to China's rising number of environmental advocates, many of whom I have had the good fortune to meet. It is through their tireless efforts that the future prospects for environmental protection in China appear so much brighter.

Abbreviations

ACCA21	Administrative Centre for China's Agenda 21
BOD	biological oxygen demand
CAS	Chinese Academy of Science
CCl_4	carbon tetrachloride
CCP	Chinese Communist Party
CCTV	China Central Television
CDM	Clean Development Mechanism
CFC	chloroflurocarbon
CICETE	China International Center for Economic and Technical Exchanges
CITES	United Nations Convention on International Trade in Endangered Species of Wild Fauna and Flora
Cn	cyanide
CO	carbon monoxide
CO_2	carbon dioxide
COD	chemical oxygen demand
CRAES	Chinese Research Academy for Environmental Sciences
Cu	copper
DAC	Development Assistance Committee (OECD)
DEAP	Dianchi Environment Action Plan
EIA	environmental impact assessment
EPA	Economic Planning Agency (Japan)
EPB	environmental protection bureau
FBIS	Foreign Broadcast Information Service, *China Daily* Report
FETC	Foreign Economic and Trade Committee
GDP	gross domestic product
GEF	Global Environment Facility
GONGO	government organized NGO
GUEPO	Guangxi Urban Environment Project Office
H_2S	hydrogen sulphide
Habitat	United Nations Centre for Human Settlements
IBRD	International Bank for Reconstruction and Development
ICSID	International Centre for Settlement of Investment Disputes
IDA	International Development Association

IFC	International Finance Corporation
ISO	International Organization for Standardization
JBIC	Japan Bank for International Cooperation
JEC	Japan Environmental Corporation
JEXIM	Japanese Export and Import Bank
JICA	Japan International Cooperation Agency
LDC	less developed country
LDP	Liberal Democratic Party (Japan)
LPG	liquid petroleum gas
LUCRPO	Liaoning Urban Construction and Renewal Project Office
MDG	Millennium Development Goals
METI	Ministry of Economic Trade and Industry (Japan)
MIGA	Multilateral Investment Guarantee Agency
MITI	Ministry of International Trade and Industry (Japan)
MoCA	Ministry of Civil Affairs (China)
MOF	Ministry of Finance (Japan)
MoFA	Ministry of Foreign Affairs (China and Japan)
MOFTEC	Ministry of Foreign Trade and Economic Cooperation (China)
MST	Ministry of Science and Technology (China)
MW	megawatt
NDRC	National Development and Reform Commission (China)
NEAPs	National Environmental Action Plans
NEDO	New Energy and Industrial Technology Department Organization
NGO	non-governmental organization
NH_3	ammonia
NO_x	nitrogen oxide
NPO	Not-for-Profit Organization
ODA	official development assistance
ODS	ozone depleting substances
OECD	Organisation for Economic Cooperation and Development
OECF	Overseas Economic Cooperation Fund (Japan)
pH	14-point scale of acidity (less than seven represents high acidity)
ppm	parts per million
RMB	Renminbi
S_2	sulphur
SAI	small area improvement
SAPROF	Special Assistance for Project Formulation Programme (JBIC)
SCP	Sustainable Cities Programme
SEPA	State Environmental Protection Administration (China)
SETC	State Economic and Trade Commission (China)
SO_2	sulphur dioxide
SOE	state-owned enterprise
SSB	State Statistical Bureau
TRAC	Target Resources for Assignment from the Core
TSP	total suspended particulates

TVE	township and village enterprise
μg/l	micrograms per litre
μg/m^3	micrograms per cubic metre
UNCED	United Nations Conference on Environment and Development
UNCHS	United Nations Centre for Human Settlements
UNDESA	United Nations Department of Economic and Social Affairs
UNDP	United Nations Development Programme
UNEP	United Nations Environment Programme
UNICEF	United Nations Children's Fund
UNIDO	United Nations Industrial Development Organization
WHO	World Health Organization
WTO	World Trade Organization
WWF	World Wide Fund for Nature
Zn	zinc

Introduction

> Whether in terms of land area or population, China is a large country which has a definite impact on the world's environment. If China's environmental problems can be solved, it will represent a major contribution to improving the quality of the world environment.[1]
>
> (Qu Geping, Administrator, National Environmental
> Protection Bureau, China)

> Nobody made a greater mistake than he who did nothing because he could only do little.
>
> (Edmund Burke)

In the Chinese municipality of Liuzhou, in Guangxi Zhuang Autonomous Region in the southwest of China, approximately 80,000 tons of sulphur dioxide (SO_2) are discharged into the atmosphere annually. Liuzhou is now one of China's leading sources of acid rain, and industry is largely to blame.[2] Located in the centre of the town, Liuzhou's Chemical Fertilizer Plant produces a tail gas of highly toxic nitrous acid known to the locals as *Huang Long* (Yellow Dragon). Established in 1966, the plant employs over 3,500 workers and makes an important contribution to local economic output. However, for over 30 years production targets have been set without any attention to the environmental consequences. The Yellow Dragon is symbolic of the serious pollution problems in China and the difficult trade-off that has to be made at the local level between economic prosperity and environmental health.

The scale and severity of environmental problems in China now threaten the economic and social foundations of its modernization. Rapid economic growth in the world's most populous nation is leading to widespread soil erosion, desertification, deforestation and the depletion of vital natural resources. All of China's major rivers and river basins are heavily polluted and it is now home to seven of the most polluted cities in the world.[3] As the world's largest producer and consumer of coal, China is also a significant contributor to global environmental stress. It produces approximately 11 per cent of global carbon dioxide (CO_2) emissions, ranking second in the world after the United States (although on a per capita basis the United States discharges over six times more CO_2 emissions than China).[4]

The regional effects of transboundary acid rain, caused by SO_2 emissions from coal combustion in China, are less well documented but potentially severe.

China's environmental problems transcend national borders and so do the solutions for dealing with them. Given the enormity of the task involved and the limited resources available in China, the international community has a critical role to play in providing financial and technical assistance. Indeed, over the past decade the Chinese government has demonstrated a stronger political commitment to environmental protection which, in turn, has attracted a dramatic rise in international development assistance.[5] Between 1993 and 1996 the total value of environmental projects funded by foreign donors in China reached US$3.2 billion – representing approximately 20 per cent of total environmental spending.[6] More recently the World Bank has provided approximately US$0.6 billion per annum in environmental-related projects representing 25 per cent of overall lending to China.[7] Japanese environmental loans to China more than doubled between 1996 and 2000 and amounted to nearly US$7.4 billion in fiscal 2000.[8] Despite an overall decline in Japanese aid to China, environmental loans have continued, reaching a total of US$4.2 billion in fiscal 2003.[9]

The overall financial contribution from international donors will always be small relative to China's environmental needs. It is, therefore, imperative that international funding is both delivered and used effectively. I argue in this book that the best measure of the effectiveness of international environmental aid is the degree to which it can strengthen capacity for environmental management, or (as it will be called henceforth) environmental capacity.

The problem is that it is by no means clear exactly how international funding and expertise can be directed towards this goal. The concept gained recognition at the United Nations Conference on Environment and Development (UNCED) in 1992. References to capacity building are scattered throughout Agenda 21 (the international plan for sustainable development). As a consequence, multilateral and bilateral donors now stress the capacity-building components of their environmental assistance programmes. Yet, little conceptual or evaluative work has been carried out to elucidate exactly what the process entails.[10] Differing interpretations range from a narrow focus on skills and training to institutional development and a broader focus on support for stakeholder participation in environmental decision-making. It may be fair to say that environmental capacity is one of the least understood areas of either environmental politics or international development. Yet, for developing countries to pursue more environmentally sustainable paths to development a focus upon capacity is essential.

Although some scholars of international relations have acknowledged the importance of environmental capacity,[11] problems over compliance and the necessary enforcement techniques have remained the central focus of inquiry.[12] In parallel to this trend at the international level, China scholars have also tended to focus upon environmental compliance rather than capacity, and very little empirical work has been done on international environmental aid to China.[13] Instead, concerns have centred upon the issues of regulatory enforcement, industry compliance and the impact of economic decentralization upon environmental reform.[14] Above all, special attention has been given to the problems associated with environmental policy implementation at the local level.

It is now widely recognized that over the past decade the Chinese government has made considerable progress in integrating environmental concerns into development policy-making. The national environmental regulatory framework is impressive relative to China's stage of development. But environmental regulations, laws and central control policies are proving difficult to implement. The limits of China's state-driven environmental management are most evident at the local level where strong evidence suggests that environmental concerns are frequently subordinated to economic imperatives. Current scholarship on environmental management in China conveys a strong impression of local non-compliance: government officials lack incentives to enforce environmental regulations, local environment agencies are politically weak, and environment officials are seen to be in collusion with the managers of polluting factories.[15]

Clearly there are limitations to what can be achieved by central fiat. Indeed, it would seem that a disjuncture exists between central consensus and local dissent that is likely to be a serious impediment to international environmental aid. For this to be effective in China, it not only requires strong political commitment at the central level but also the political support of local provincial and municipal agencies that are responsible for the implementation and repayment of environmental loans.

The question that needs to be addressed is whether local governments in China are uniformly opposed to environmental reform. Although weak compliance is clearly a severe impediment, weak capacities, including low levels of technological expertise, weak institutional coordination, and poor financial management (to name but a few) are also likely to pose a serious constraint. In some cases, strengthening capacity could be a necessary pre-condition for compliance rather than the reverse.

Developing environmental capacity in China

This book challenges the prevailing wisdom that weak compliance is the only constraint upon local environmental management in China. It advances two interrelated discussions. First, it seeks to construct a conceptual framework for understanding the key dimensions of environmental capacity. This is broadly defined to encompass the financial, institutional, technological and social aspects of environmental management. Within these broad dimensions, I argue that the test of environmental assistance must be the degree to which it is able to stimulate self-reliance on the part of the beneficiaries involved. In other words, environmental capacity building must endure over time.

Second, the book presents the results of an empirical inquiry into the implementation of donor-funded environmental projects in both China's poorer and relatively developed regions. By drawing upon extensive fieldwork, it seeks to explain how, and under what conditions, international donors can strengthen China's environmental capacity, especially at the local (provincial, municipal and county) level.[16] The focus of the investigation is on urban environmental problems – the so-called 'brown issues' of air and water pollution and solid waste. This reflects China's immediate environmental priorities and, as a consequence, the dominant direction of current environmental aid flows. Ultimately, the book aims to expand the debate on environmental management in China by providing a more complex picture of the difficulties experienced at the local level and by linking this experience to international aid efforts.

A pluralistic approach

A focus upon environmental capacity building as the key condition for under-standing donor effectiveness is a complex task. There exist few systematic accounts of the actual impact of environmental assistance upon capacity building. In general, scholars have been overly concerned with the most appropriate means of managing environmental problems rather than the necessary pre-conditions and capacities for doing so. It is widely recognized that the capacity to implement environmental goals depends in part upon effective bureaucracies and state regulatory practices. More recently, scholars have focused upon the need for efficient market tools and participatory practices as the prerequisites for sustainable environmental management. The problem is that theorists have tended to promote a single vision – state-centric, market centric or community centric – rather than a synthesis between them. In addition to the tendency towards mono-causal approaches, a huge lacuna exists with respect to understanding how market or participatory approaches actually affect capacity building in developing countries.

By analysing the relationship between different approaches to environmental management and local capacity, this book makes two interlinking arguments. The first is that a pluralistic approach to managing environmental problems is an essential means of building capacity. Contrary to the main assumptions underly-ing the theoretical debate on environmental management, in practice there is no conflict between advocating economic incentives and supporting participatory practices. Both approaches provide a complementary means of building environ-mental capacity. Simply put, economic incentives can improve financial and institutional efficiency and participatory practices can facilitate institutional cohesion and technological innovation. They are equally important for securing political commitment and, combined with state regulation, have a greater potential to overcome economic power asymmetries and weak compliance at the local community level.

The second argument is that for environmental capacity to endure over time we need to pay more attention to the social dimension. Scholars and practitioners alike have tended to focus solely upon the instrumental dimensions of environ-mental capacity. As later chapters will reveal, new information, institutions and technologies are important determinants of environmental capacity but they are unlikely to prove effective without a normative shift in thinking and behaviour on the part of the actors involved. Above all, environmental capacity relies upon a strong sense of shared responsibility amongst government agencies, industries and citizens. This is especially important in the case of China where a predominantly top-down approach to environmental management has reinforced public perceptions that the environment is essentially a government responsibility. Psychological resistance to the notion of shared responsibility is likely to be strong because in the current era of capitalism and individual incentives it appears to be regressive and reminiscent of the old commune system.

The question of how to overcome this inertia brings us back to the first argument and the importance of a pluralistic approach. In an authoritarian context the need for greater public participation is obvious. And as this book reveals, there are some signs that the Chinese government is moving in this direction. However, participation per se is not enough. It plays an important part in nurturing a sense of responsibility but economic incentives are also needed to ensure that commitment persists over time. The co-dependency between the two approaches lies in the fact that effective environmental capacity depends upon the dual principles of shared responsibility and economic viability.

Case studies

In addressing the issue of aid effectiveness, it is also necessary to take into account the diverse nature of environmental aid to China: not only are there a large number of donors involved but each donor has its own priorities, approach and principles with respect to environmental management. By and large, the three main donor approaches in China focus upon building environmental infrastructure, introducing market measures, and strengthening environmental decision-making on the basis of stakeholder participation. An initial expectation drawn from the scholarship on China suggests that both market-oriented and participatory approaches to environmental management would be difficult to implement because of the transitional nature of the Chinese economy and the authoritarian nature of state–society relations. In attempting to solve its environmental problems, the Chinese government places a far stronger emphasis upon regulatory control than market forces or various forms of participation. Moreover, the traditional Chinese approach to environmental management is highly technocentric, based upon the underlying premise that environmental problems can best be solved through improved technologies and engineering capability. Under such circumstances, how far can donors move beyond the need for basic environmental infrastructure in China? How important is building technical competence relative to upholding sound environmental management principles? Is the introduction of a market incentive approach more likely to facilitate environmental capacity than a participatory approach? How do different donor approaches affect local political commitment and engagement? And under what conditions can different donor approaches to environmental management strengthen local capacities?

To address these questions, this book adopts a comparative case study approach by focusing upon three major environmental donors in China – the World Bank, the United Nations Development Programme (UNDP) and Japan. The World Bank is an obvious focus of attention given its dominant financial and advisory role in

China. The UNDP merits attention for its widely accepted role as the central international proponent of sustainable development. Amongst the many bilateral environmental donors, Japan stands out for a variety of reasons. Japan leads the Organisation for Economic Cooperation and Development (OECD) countries in providing environmental aid and China is a major recipient. It is now globally competitive in clean technologies with particular experience in dealing with industrial pollution. And finally, it is located downwind from one of the world's largest concentrations of coal-fired power plants. Unlike CO_2 emissions and global warming, SO_2 producing acid rain is visible in the immediate term and in the case of China and Japan affects both polluter and victim.

All three donors combine a number of priorities in their environmental aid packages to China relating to infrastructure building, poverty alleviation, institution building and skills training. However, the three donors have adopted sharply contrasting approaches to environmental management: Japan focuses its attention upon engineering efficiency and technical competence; the UNDP stresses the importance of human development with an emphasis on participatory practices and technological advance; and the World Bank lays emphasis upon the principle of appropriate price reform and institution building (see Table 0.1). These three donor approaches – *engineering* (Japan), *human development* (UNDP) and *market-institutional* (World Bank) are broadly representative of the dominant forms of environmental assistance in China; the human development and market-institutional approaches also reflect the key principles underpinning the theoretical debate on environmental management.

It is important to stress here that the main concern of this study is with the variation in donor approaches to environmental management and their effects upon local environmental capacity regardless of differing interpretations and practices

Table 0.1 Comparative donor approaches to environmental management in China

Approaches	Japan	UNDP	World Bank
Engineering approach: degree to which the design and implementation of projects are solely determined on the basis of engineering practices	High	Low	Medium
Human development approach: degree to which participatory practices and technological advance are integrated into the design and implementation of projects	Low	High	Low
Market-institutional approach: degree to which the design and implementation of projects are conditional upon utility pricing and institutional reform	Low	Low	High

with respect to capacity building. With the exception of the UNDP's programmes, very little environmental assistance to China has focused specifically upon capacity building. With many environmental projects, capacity building is not the main intention of the donor. Nevertheless, this does not invalidate an investigation into any possible indirect effects or unintended consequences. Moreover, an analysis of what donors should be doing to promote environmental capacity building is both timely and necessary.

For the Japanese, capacity building equates with 'self-help'. This essentially means that local beneficiaries should rely on their own resources, with small amounts of funding to stimulate scientific and technological advances. This has its modern equivalent in the form of 'recipient ownership'. Self-help is a Victorian concept that was imported to Japan during the Meiji restoration and vigorously applied to its economic reconstruction after the Second World War. Unlike Europe, Japan did not receive reconstruction funds in the form of a Marshall Aid Plan and, therefore, largely had to rely on its own resources.[17] Anecdotal evidence suggests that this experience has left an indelible mark on the Japanese way of thinking with respect to development assistance.[18] In practice, the capacity building component of Japanese environmental assistance is relatively small and narrowly focused upon enhancing human skills and training.

In stark contrast, capacity building is a crucial element of the UNDP's central concern with human development. Following the UNCED in 1992, the UNDP launched the Capacity 21 Fund which involves assisting developing countries with integrating environmental imperatives into their national plans for economic development. On this basis the UNDP has made some progress in broadening the concept of capacity building to include institutional, technological and informational dimensions. The World Bank also stresses the capacity building component of its projects and programmes but it has a less targeted approach than the UNDP. It essentially focuses upon institutional development that involves the introduction of economic incentives, the development of market-oriented institutions and the decentralization of environmental management. It also places some emphasis upon building human resources.

The three donors also differ in the types of environmental assistance that they provide. As a bilateral donor, Japan's environmental assistance to China is constrained by the nature of the Sino-Japanese relationship. The UNDP and the World Bank, as multilateral donors, have more room to shape their own environmental agendas, although China's power status in the world means that donor leverage is weak relative to other developing countries. Moreover, the UNDP's environmental assistance is grant-based, whereas the World Bank and Japan also provide environmental loans. These differences are not central to the comparative dimension of the study. Even so, they cannot be ignored and shall be introduced into the analysis where relevant.

The environmental projects

All of the 13 projects selected in this study are broadly representative of each donor's environmental assistance efforts in China. In the cases of Japan and the

World Bank the focus is upon environmental loans rather than environmental grants because they are specifically related to environmental pollution. The projects have been selected on the basis of three key criteria. First, as outlined earlier, the key variance across all 13 projects lies in the environmental management focus of the donor – *infrastructural, human development* or *market-institutional*. Second, the project sites are located in China's most polluted cities which have also, with the notable exceptions of Guiyang and Huhhot, received relatively high concentrations of international environmental assistance.

Third, to provide a comparative regional perspective, all the project sites are located either in China's poorer regions (north and southwest) or its relatively developed regions (central and northeast) (see Table 0.2). Differing conditions within these regions provide a means for testing the validity of uniform donor solutions. For example, in the industrialized northeast, the slow reform of state-owned industry poses the predominant challenge to environmental management while in the poorer southwest agricultural malpractice still remains a central concern. Moreover, in the northeast and southwest regions of China the local capacity to solve environmental problems is weak in contrast to the more economically advanced coastal regions.

The aggregate cost of the projects amounts to approximately US$3.5 billion. Of this total amount the three donors have provided US$1.62 billion (US$1.3 billion from Japan, US$0.3 billion from the World Bank and US$1.9 million from the UNDP) with the remaining costs funded by local governments (provincial, municipal and county) and enterprises (see Map 0.1 for an overview of the project sites).

It is important to stress here that international environmental aid to China has only gathered momentum since the mid-1990s. Consequently, in this study any

Table 0.2 Regional location of donor-funded environmental projects in China

Donors	Poorer regions	Relatively developed regions
Japan	Liuzhou environmental improvement project	Shenyang environmental improvement project
	Guiyang model city project	Benxi environmental improvement project
	Huhhot environmental improvement project	Dalian model city project
UNDP	Capacity building for acid rain and SO$_2$ pollution control in Guiyang	Managing sustainable development in Shenyang
		Managing sustainable development in Wuhan
		Capacity building for the widespread adoption of cleaner production for air pollution control in Benxi
World Bank	Yunnan urban environmental project	Liaoning urban environmental project
	Guangxi urban environmental project	

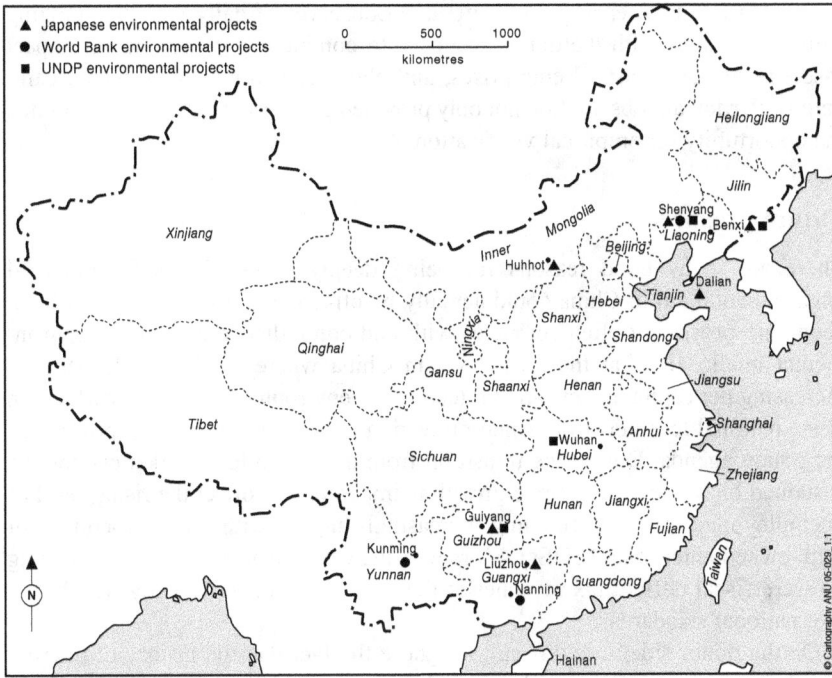

Map 0.1 Location of selected environmental projects in China.

Source: National Geographic, *Atlas of the World*, 5th edn, Washington, DC: National Geographic Society, 1981.

attempt to evaluate the full environmental impact of donor-funded projects would have been premature and quite possibly misleading. At the initial stage, it seemed that the study would be greatly impeded by the fact that the selected environmental projects were still ongoing. It was not possible to organize the research around any expected completion dates because, for either political or economic reasons, long time delays could be expected at any stage of the implementation process. How could one evaluate a project that had not been completed? But as the research progressed it soon became clear that observing these projects while still 'alive' was critical to uncovering the complexities of the interactions between the donors and the local agencies and to assessing any changing attitudes on the part of the beneficiaries involved. In the cases where projects have been completed, I have made references to the findings of the donor project evaluations. It is important to note, however, that the purpose of these evaluations is to gauge the extent to which the projects have met their stated objectives. They are therefore a poor indicator of the actual impact of any particular project upon local capacity.

Aside from relying upon interviews, environmental surveys and factory visits, my main method of evaluation used at the local level was to observe 'a day in the life of an environment official'. I spent up to a week with environmental protection bureaux officials at each project site as an observer of daily activities ranging from discussions with the donor agencies, to conducting environmental impact assessments for industrial enterprises, and planning and managing public campaigns. Participant observation not only provided greater insights but also granted the opportunity for empirical verification.[19]

Future prospects

There are many good reasons for being deeply sceptical that international environmental aid to China could actually be effective on the ground. In China, economic prosperity often competes with and contradicts the need for environmental quality. It is at the local level in China where the trade-offs between increasing per capita income and protecting the environment are most stark. Even at the national level, environmental protection is by no means the top priority on the policy agenda. During the transition from a planned to a market economy a sustained high rate of economic growth is imperative to offset the rising level of unemployment. This is becoming increasingly urgent with the restructuring of state-owned enterprises (SOEs). Moreover, the Chinese economy is experiencing a severe fiscal deficit; tax revenues at the central level are currently well below international standards.

On the donor side, it is difficult to ignore the fact that the history of international development assistance has been beset by complex problems relating inter alia to administrative ineptitude, donor self-interest and the strategic behaviour of recipient governments.[20] The environmental record of international donors, especially Japan and the World Bank, has been dismal. In the Japanese case, criticism has centred upon Japan's consistent and large-scale usage of official development assistance (ODA) funds for infrastructural development in Asia at a substantial cost to the environment.[21] Historically, 60–70 per cent of ODA has been spent on infrastructure, cynically referred to as *hakomono enjo* (assistance for construction).[22] The World Bank has also been heavily criticized for its predominant focus upon economic growth at the expense of the environment. Environmental projects have often been excessively costly while at the same time producing unfavourable social and environmental consequences.[23]

But against all odds, the findings in this book provide some grounds for optimism. First, non-compliance at the local level in China is not the only constraint upon effective environmental management; weak capacities also pose a serious problem. In many ways, given their proximity to the environmental problems at hand, local environment officials are more committed to the task of protecting the environment than their counterparts in Beijing. This is contrary to what many China scholars working on environmental issues would predict.

Second, at a broader level, it is clear that China is moving away, albeit slowly, from its traditional state-centric focus on environmental management and towards a more comprehensive paradigm that embraces both market solutions and to a lesser extent participatory practices. Despite political and structural constraints, environmental donors have had a small but significant part to play in facilitating this shift. Third, the study demonstrates that under certain conditions international environmental aid has had some positive effects upon environmental capacity in China. Taking into account local economic, political and spatial conditions could further enhance its effectiveness. At the same time, more effort is needed to overcome a number of weaknesses on the part of the donor institutions, especially in relation to a low level of commitment, institutional inflexibility and weak inter-donor coordination.

These conclusions have important theoretical and policy implications. At the theoretical level, they suggest that the ideological divide that exists between market proponents and participatory advocates is largely redundant. Even in an authoritarian state, such as China, that is experiencing massive economic and social upheaval, market-oriented and participatory approaches are both relevant to enhancing environmental management and are most effective when applied in combination. This is not to imply that either approach could act as a substitute for state action, only as a counterbalance. In authoritarian settings, although it is tempting to advocate the retreat of the state and the subsequent takeover of non-governmental forms of environmental management, it is important to acknowledge that the state still has a central role to play in facilitating environmental cooperation between government agencies, industries and citizens.

From a policy perspective, the findings highlight the urgent need to balance efforts towards ensuring environmental compliance with efforts to build environmental capacity. For officials in Beijing it is clearly easier to place the blame for China's weak environmental management upon non-compliance at the local level rather than address local capacity constraints. Likewise, it is far easier for environmental donors to focus solely upon infrastructure building or pricing reform rather than the more complex issue of capacity building. Yet, if significant progress is to be made in solving China's looming environmental crisis, then developing local environmental capacity needs to be placed firmly at the top of the policy agenda. In order to redirect development assistance towards that goal both donors and counterpart agencies in China need to undertake considerable institutional adjustments. Above all, institutional practices must be driven by concerns for effective outcomes at the local level.

To end on a more cautionary note, although in this book the case for developing environmental capacity is made unequivocally, one of the major lessons from this study presents a dilemma. For environmental capacity building to fully succeed, a certain level of pre-existing capacity is also required. The uneven spatial distribution of China's economic growth is leading to a situation in which the ability to act environmentally is higher in the more developed regions where pre-existing local government financial and institutional capacities are higher. Concentrating on China's richer regions, however, would run counter to the international donor community's overriding goal of poverty alleviation. Moreover, it would result in resources being diverted away from those regions in China where environmental problems are the most severe. Tackling this dilemma, which involves identifying the difficult causal linkages between poverty and environmental degradation, remains the single biggest challenge for international environmental donors currently working in China.

This book is organized around the central theme of environmental capacity. Two key sub-themes relating to 'incentives' and 'participation' also form common threads throughout the six chapters. Chapter 1 discusses the strengths and weaknesses of both market and participatory approaches to solving environmental problems. By taking into account implementation concerns, especially in developing countries, it stresses the need for a pluralistic approach to environmental management that combines markets with participatory practices and state regulation. The chapter also discusses the various interpretations of capacity building and provides a conceptual framework for understanding the key dimensions of environmental capacity.

Chapter 2 provides the historic and contextual background for the three case studies. It sketches the nature and scale of environmental problems in China together with the central government response. Attention is given to assessing China's evolving state-centric system of environmental management and to identifying the problems involved in implementation at the local level. It concludes by considering the implications for international environmental assistance. The chapters that follow comprise the three case studies. Chapter 3 deals with the Japanese approach to solving environmental problems in China, Chapter 4 with the UNDP approach and Chapter 5 with the World Bank approach. All three chapters explore the interests of donors, the project implementation process, local responses and the contribution of each donor approach to strengthening environmental capacity. Chapter 6 provides a detailed comparative analysis of the major empirical findings and further refines and develops the conceptual framework. Special attention is given to emphasising the necessary conditions for building environmental capacity in China and to the issue of sustainability over time. In addition, the chapter identifies specific constraints on the part of the donor institutions. The book concludes by formulating some lessons for the future development of environmental capacity in China.

1 Developing environmental capacity

> What is common to the greatest number has the least care bestowed upon it.
> Everyone thinks chiefly of his own interest, hardly at all of the common interest.
>
> (Aristotle, *Politics*, Book II, chapter 3)

> A people is not any collection of human beings brought together in any sort of
> way, but an assemblage of people in large numbers associated in an agreement
> with respect to justice and a partnership for the common good. The first cause of
> such an association is not so much the weakness of the individual as a certain
> social spirit which nature has implanted in man.
>
> (Marcus Tullius Cicero, *The Republic*)

Over the past two decades the environment has attracted widespread international attention. National governments, in both developed and developing states, now recognize the importance of protecting the environment in order to sustain economic and social development, multinational corporations are cognizant of the need to reduce the negative impact of their products and services, and rising numbers of transnational environmental movements are forming to activate public awareness. Yet, despite the growing global consensus over the need for environmental protection, opinion is still deeply divided over how best to achieve it.

The theoretical discourse on environmental management is primarily concerned with ways of influencing behaviour. An underlying assumption is that behavioural change will ultimately lead to fewer environmental problems. In addressing this goal, however, a clear ideological divide exists between proponents of market solutions and advocates of public participation that diverts attention away from the linkages between the two perspectives, as well as important implementation concerns. Supporters of both these approaches have given only limited attention to the critical question of how to induce behavioural change in developing countries such as China where conflicts over environmental and economic priorities are intense and where social organization is more heavily controlled by the state. In such contexts, environmental problems need to be analysed on the basis of how to strengthen environmental capacity as well as influence behavioural change.

But scholars have largely overlooked the question of how diverse approaches can actually strengthen capacity to pursue environmental goals more effectively.

This gap in understanding has important implications for the delivery of international development assistance. The new discourses on environmental economics, sustainable development and civil society have prompted international donors to incorporate economic incentives and participatory practices onto their environmental aid agendas without a clear idea of the necessary conditions for sustaining such approaches. Within the donor community there appears to be a *prima facie* assumption that sound environmental principles will inevitably lead to effective solutions.

This chapter explores the relationship between different approaches to environmental management and the nature of environmental capacity. It is divided into two parts. The first part advocates a pluralistic approach to environmental management that combines economic incentives with participatory practices and state regulation. The second part focuses upon capacity building. It draws upon various interpretations of the concept and proposes a conceptual framework for understanding the broad dimensions of environmental capacity. The actual interplay between divergent environmental approaches and capacity needs to be determined on the basis of empirical investigation. The purpose of this chapter, therefore, is not to elucidate the linkages but rather to provide a framework of analysis for interpreting the case studies presented in Chapters 3–5.

A pluralistic approach to environmental management

Traditional theory on environmental management is rooted in the fundamental premise that self-interested behaviour will inevitably lead to environmental degradation. In the 1960s, based on work such as Mancur Olson's collective action thesis (whereby individuals will not act to achieve their common or group interests)[1] and Garrett Hardin's metaphor of the 'tragedy of the commons',[2] theorists proposed a stronger role for state intervention.[3] These theories were later contested on the basis of a wealth of empirical research into state-induced environmental failure. In the 1980s and early 1990s, the state almost retreated from sight in the wake of neoliberal economics and public-choice theory. The self-regulatory function of the market was deemed to be superior to the command and control function of governments. The platonic view of government was also seriously tempered by scholars who stressed the importance of non-governmental organizations (NGOs). They advocated decentralization and empowerment in facilitating effective environmental management and the need for grassroots pressure to reform the state.[4]

There are at least three significant problems with a state-centric approach to environmental management. First, the environment is complex, involving multiple causal relationships and a high degree of what environmentalists refer to as 'problem displacement' (i.e. the solution to one environmental problem can generate another problem). For example, electric cars could be introduced to solve traffic induced air pollution but this in turn would require more power plants in order to satisfy the increase in demand for electric power. The question then becomes which would be more harmful to the environment, power plants or automobiles? Governments and scientists alone cannot determine the exact requirements and scope of environmental protection – information is partial and monitoring and transaction costs are high.

The second problem is that state-led environmental regulation suffers from an 'implementation deficit'. As many economists predict, the implementation of environmental goals is often severely constrained by special interest groups and self-seeking officials. Third, even if one assumes that not all local officials are corrupt, environmental protection from above can undermine local commitment and responsibility towards the environment. Above all, it lacks democratic consent.

In response to the insufficiencies of the state, a more contemporary debate has emerged over the roles of 'markets' or 'citizens'. Two leading schools of thought have developed in relation to economic incentive and participatory approaches to environmental management. Essentially, the market-oriented approach advocates proper pricing policies and the use of economic instruments, whereas the partic-ipatory approach stresses the need for information exchange and stakeholder participation. These competing theories are in part ideologically driven but they also share an understanding of the limitations of the state in solving environmental problems.

The market approach

Traditionally the relationship between economic development and environmental protection has been seen as antagonistic: development per se was interpreted as the root cause of environmental degradation and environmental protection was interpreted as the antithesis of economic growth. In the 1970s, it appeared that the economic growth proponents had won the battle. The Club of Rome (1972) thesis on 'Limits to Growth' (that the world would face ecological collapse if current trends in industrialization and population growth persisted) met with severe criticism because, despite such catastrophic predictions, rising living standards continued throughout the developed and developing world. Essentially, the Club of Rome demonstrated the weakness of extrapolation analyses that do not adequately take into account incremental technological change.[5]

However, since the 1980s the discourse on the environment and development has shifted towards reconciliation in response to new scientific knowledge, a number of environmental disasters and growing public concerns. Rather than an impediment to growth, environmental conservation and protection is now seen as a necessary prerequisite for long-term economic development because it can, in turn, lead to enhanced economic efficiency and technological innovation.[6] In particular, environmental economists have had a major influence upon the policy discourse by redefining markets to meet the requirements of a more environmentally sensitive world.

Environmental economics is based upon mainstream economic analysis of pollution problems in free market systems.[7] This approach dates back to the work of Arthur Pigou in the 1920s.[8] It is concerned with the question of externalities – spillover effects borne by those not party to an economic transaction (as in, for example, the dumping of toxic waste leading to the contamination of a whole community's water supply). To correct for externalities, environmental economists advocate the need for *getting the prices right* by abolishing subsidies which do not reflect the marginal costs of production, and incorporating the full social costs (i.e. environmental externalities) into the price mechanism. To this end, economic instruments such as charges (raising revenue for abatement costs) and taxes are proposed.

However, the question of how to calculate green taxes and charges remains a contentious issue. Many environmental economists would argue that attaching monetary value to environmental goods is essentially the responsibility of the government.[9] What is crucial is that charges and taxes must be sufficient to provide a disincentive against polluting the environment, or over-consuming natural resources, and instead encourage technical change. Recent market-based instruments are more sophisticated – taking the valuation problem into account by suggesting more innovative means such as tradable permits (whereby governments set the ambient level of pollution or resource extraction rate and tradable permits allow for the allocation of emissions or resources below the prescribed threshold).[10]

The important point to make here is that economic instruments are not solely dependent upon markets. Clearly, any pure market approach is inadequate; profits will always outweigh environmental concerns. Instead, economic instruments and incentives are a means of increasing the efficiency of pre-existing environmental regulation and providing both governments and firms with greater flexibility.[11] The same, of course, can be said for other market-based instruments such as tradable permits and deposit-refund schemes.

In practice, although environmental regulation is still the dominant norm amongst the OECD countries, economic instruments are gaining increasing popularity in Europe and tradable permits are winning support in the United States. In the case of developing countries, the market approach remains largely untested, and the question of whether it should be promoted has caused a major rift between environmentalists and economists. Many environmentalists are ideologically opposed to the very idea of 'internalizing' environmental costs through the market mechanism in developing countries; it is seen as premature given the low levels of per capita incomes and education. In the words of Michael Redclift, 'on its own, resource accounting tacitly endorses a highly technocentric and North-biased view of the development process'.[12] In contrast, economists such as Theodore Panayotou argue that economic instruments will benefit developing countries by reducing rent-seeking behaviour (i.e. by reducing the opportunities for collusion between the regulators and the regulated) and of generating revenue.[13]

The problem is that in both developed and developing countries, economic instruments are mainly used as a source of revenue. It is difficult to impose them at sufficiently high levels because of industrial concerns over the rising costs of production and a consequent reduction in international competitiveness. Green taxes, therefore, provide a weak incentive for firms to modify behaviour that is consistent with good environmental practice. Moreover, tradable permit schemes do not sufficiently 'stigmatize' environmental pollution; they are often perceived as a 'license to pollute'. Robert Goodin refers to tradable permits as 'selling environmental indulgences' – similar to the medieval Catholic Church selling indulgences to sinners during the crusades.[14]

Like their more orthodox predecessors, environmental economists still have a tendency to reify the market without giving due consideration to the fact that environmental problems are caused as much by social practices or patterns of consumption as competitive modes of production. Incentives, therefore, cannot obviate the need for collective action. In general, environmental economists show only a limited understanding of the need for individual citizen or community participation in environmental management. Michael Jacobs argues the point quite persuasively:

> [The] market makes a good servant but a bad master. It can be used, through a variety of financial incentives, to help society achieve its objectives. But these objectives should be chosen through a public, democratic process, not a private, market one. The environment is (most of the time) a public good; its protection or consumption should therefore be a matter of social concern, and collective action.[15]

The participatory approach

The participatory approach offers a critique of both economic reasoning and state involvement in environmental affairs. In contrast to the positivist approach of evaluating individual preferences and then designing incentives to change behaviour, participatory proponents emphasize the need for a process of dialogue to encourage convergence over perceptions of the public interest. Participation is believed to lead to a change in behaviour on the basis of the exchange of information and negotiation over preferences.

The approach stresses the importance of citizenship and 'the social spirit'. The citizen is not seen as an alternative to the rational self-interested being. Rather, both citizenship and self-interest are perceived to coexist. The environmental philosopher Mark Sagoff, for example, makes the distinction between consumers seeking to maximize their individual preferences according to their personal welfare concerns, and citizens who act on community values derived from conceptions of 'national pride and collective self-respect'.[16] Such philosophical reasoning also underpins the arguments against centralized environmental decision-making. Many theorists have argued that without public involvement, environmental management is doomed to fail because effective implementation relies upon public consultation and engagement. They reflect the argument advanced by the American ecologist Barry Commoner – that external constraints such as the market and the state will only work if they are ubiquitous.[17]

Participation can be interpreted in a number of ways ranging from representative elections to grassroots women's cooperatives. We need not, in the present context, examine every aspect of the possible forms of participation. For our purposes, the concept of participation can be understood more easily if we distinguish between two dominant perspectives within the literature – the *democratic perspective* rooted in the civic republican tradition and the *development perspective* rooted in the communitarian tradition. Both positions advocate the decentralization of power, public dialogue and voluntary action, but they differ in their assumptions over the relationship between the state and civil society.

The democratic perspective advanced by political theorists stresses the need for open public discussion among the different stakeholders within society.[18] It does not necessarily involve a transfer of power from the state to its citizenry. Public participation can simply mean stronger consultation between government and citizens. In this interpretation, participation is more than an instrumental means of achieving certain policy goals; it is an intrinsic good in itself.[19] Political theorists also define participation in relation to societal activism. Drawing upon the work of Alexis de Tocqueville, associational activity in neighbourhood associations, charity organizations, schools and sports club activities is seen as a more effective means of achieving environmental objectives than bureaucratic modes of social organization. Environmental social movements are a good example of how voluntary association can lead to effective policy outcomes, especially at the local level.

As noted by many commentators, one problem with the democratic perspective is that it fails to take into account the existence of powerful interest groups either within the government bureaucracy or in large corporations. A debate among equals it is not: interested parties such as community representatives and environmental groups can easily be consulted and then just as easily be ignored; debate, in this instance, is merely an exercise in political legitimacy.[20]

In a more pragmatic vein, participation also raises problems of scale, complexity and time. Historically, participation or deliberative democracy is an Athenian ideal that was practised on the basis of face-to-face consultation. In relation to the modern ideal of representative democracy, small-group debates or stakeholder panels provide an alternative solution. But even small groups cannot avoid the time-consuming nature of debate as recognized in the famous adage by Oscar Wilde: 'the problem with socialism is that it takes up too many evenings'.

From an environmental perspective, it is also important to acknowledge that democracy is no easy remedy. Cross-national studies carried out by the Research Unit for Environmental Policy at the Free University of Berlin confirm that democratic institutions on their own seem to play a minor role in environmental policy.[21] The environment literature on the former Soviet Union and Eastern Europe is sobering in this respect. Many commentators focused upon the lack of citizen participation in policy formulation and implementation as the key systemic constraint upon environmental protection in the former Soviet Union.[22] In the wake of the Chernobyl incident in 1986 the political system allowed the then President Mikhail Gorbachev to orchestrate an impromptu 'greening' of domestic and foreign policy. Glasnost provided the necessary opening for public involvement. But environmental movements were more effective in pushing for political outcomes (i.e. the fall of Communism) than environmental outcomes. Post-1989, analysts were forced to acknowledge that following the collapse of

Communism new environmental movements were severely undermined by conflicting economic priorities.

The development perspective on participation places less emphasis upon democratic reform at the state level and more emphasis upon the need for 'bottom up' environmental management.[23] The devolution of power from the state to the community is central, together with a strong focus upon 'local knowledge' (in the belief that local people can contribute important ideas, technologies and organizational capabilities to the development process).[24] Local community participation is considered to be an integral part of environmental management: local people have a greater stake in protecting the resources upon which their livelihoods depend; local resources are easier to mobilize; and local supervision and information exchange can reduce transaction costs considerably.

However, the emphasis upon 'community spirit' tends to mask the differences in political and economic power within communities. As Lore Ruttan observes 'this is even more problematic today given the manner in which changing economic opportunities, particularly differential access to capital, are rapidly altering the balance of power in many communities'.[25] The implementation of a participatory approach raises a particular problem for authoritarian states because it assumes a democratic setting. Community participatory planning in practice takes place largely in countries with a strong local democratic tradition such as the Philippines, Sri Lanka and India.[26] In India, community participation is beginning to succeed despite the subordination of local institutions to the political organizations of the party and the state. But success is less apparent in authoritarian states with strong political hierarchies.

Clearly no single panacea exists for managing environmental problems. The strength of the market-oriented approach lies in providing efficiency gains through the utilization of financial incentives. Evidence suggests, however, that pollution charges or environmental taxes are unlikely to be set at a sufficient enough level to achieve the desired environmental improvement. Instead, incentives work best when combined with regulation and participation. For example, high water prices are more effective in reducing consumption when combined with regulatory standards and public campaigns to save water. Equally, the participatory approach cannot guarantee that democratic consultation or community-led cooperation will undoubtedly lead to better environmental outcomes. Let us now explore an alternative approach that is more aligned to striking a balance between economic incentives and participatory practices. It has particular relevance to those countries such as China that are striving to control pollution in complex urban settings as well as traditional rural communities.

A middle ground

The European body of ideas on ecological modernization[27] focuses upon environmental pollution problems and stresses the need for partnerships between governments, industries and citizens.[28] A central proposition is that the political economy should be restructured to account for environmental quality. At the macro level, structural change is needed to shift the economy away from energy- and resource-intensive industries and towards service and knowledge-based industries. At the micro level, advocates propose moving away from *end-of-pipe solutions* (i.e. add on products and processes which limit the total amount of emissions) and towards *cleaner production* (i.e. technologies and systems which are directly integrated into the design and operation of industrial production and which treat environmental pollution at source).[29]

Ecological modernists argue that future economic development will be linked to higher standards of pollution control and a rising demand for clean products. Economic incentives are seen as a necessary means of encouraging industry compliance but not sufficient to convince industry of the economic benefits of environmental protection. Hence, developing partnerships between government, industry, science and mainstream environmentalists is deemed essential. In the words of Andrew Gouldson and Joseph Murphy, a 'hands on approach to imple- mentation' is needed, which involves 'expert regulators interacting with regulated companies in a flexible, intense and cooperative way'.[30] This approach, however, does not escape the fundamental problem with the participatory approach: that improved interaction between stakeholders does not necessarily lead to better outcomes; interaction can also lead to co-optation or even corruption.

In its conservative form, ecological modernization theory is merely concerned with the 'retooling of industry'.[31] At the other extreme, it is theoretically embedded in debates about modernity. The radical variant of ecological modernization provides a sharp departure away from mainstream democratic participation as discussed earlier – decision-making involves citizens not only in respect to micro- level public investment concerns but more broadly in relation to the nature and trajectory of economic development. This interpretation draws upon the work of Ulrich Beck's theory of the 'risk society' together with his notion of 'reflexive modernization' proposed in collaboration with Anthony Giddens.[32]

In brief, the risk society perceives modern society as 'contingent, ambivalent and (involuntarily) open to political arrangement'[33] rather than the product of technical and bureaucratic rationality. The fundamental tenet of 'reflexive modernization' is that in the age of global risks it is futile to rely upon the political philosophies, models and methods of the nineteenth century industrial society. Instead, with public perceptions of risk as the necessary foundation, decision- making needs to become less expert-centred and depoliticized. In short, issues over the consequences, and not only the promise, of technical change and economic development need to be subject to public scrutiny.[34]

The radical variant of ecological modernization is an interesting proposition but it remains vague in practice. How exactly can a transformation of the industrial economy take place? How can a modern society based upon critical self-awareness rather than expert knowledge overcome traditional institutional barriers and ways of thinking? And, perhaps more importantly, how can we be certain that public consensus rather than expert advice will produce better outcomes in relation to the environment? Arthur Mol observes, insofar as environmental issues are still defined along the lines of pollution control and material throughput, that there appear few prospects for a departure from the technocratic and corporatist problem-solving approach.[35] The conservative version of ecological modernization, therefore, is more likely to prevail.

Bringing the state back in

A central problem with all three theoretical perspectives outlined earlier is that they fail to give adequate attention to deep-seated political and economic power structures. They also advocate minimal intervention from the state. Given our concern with China, the question that needs to be addressed here is how do these theories apply to those countries in which the state has ultimate political authority, where markets are still in the transition phase from central planning, where democracy is still remote and where arguably the need for environmental management is greatest? If non-state approaches to environmental management are to succeed in developing and transitional contexts, they will still require strong political commitment from the state. As discussed earlier, market instruments cannot function without the state, which is responsible for either attaching monetary value to environmental goods and services (as in the case of environmental taxes) or determining the ambient level of pollution (as in the case of tradable permits). It would also be unrealistic to assume that participatory practices alone could persuade citizens to prioritize environmental concerns over development needs. State regulations are equally important. Even a corporate focus on industrial restructuring and technological innovation is unlikely to succeed without the support of state subsidies or investment.

A pluralistic approach to environmental management that combines non-state practices (market incentives and citizen participation) with state-led regulations has a stronger potential to influence behavioural change. However, the political, economic and social constraints involved in implementing such an approach in developing countries cannot be ignored. Simply advocating diversity is no guarantee that any positive changes in behaviour will endure over time. For this, we need to understand how particular approaches actually build environmental capacity. We shall be seeking to elucidate some of the complexity of the relationship between approaches and capacity in the detailed empirical studies in Chapters 3–5. Here it is necessary to arrive at a better conceptual understanding of what environmental capacity entails before we can establish how it is affected by various approaches.

Conceptualizing environmental capacity

Conceptual difficulties abound in thinking about the idea of environmental capacity, not least because the concept has been largely ignored within the literature. Although much has been written on the subject of capacity building, most of the literature employs a narrowly defined administrative definition that is too limited for use in relation to the environment. A focus upon skills and training cannot capture the broader implementation constraints upon effective environmental management in developing countries.

The notion of capacity building is rooted in the institution-building literature of the 1960s and 1970s, which focused upon planned and guided social change within formal organizations.[36] For many, institution building was synonymous with political development or modernization.[37] For others, institution building was process driven: institutionalization was seen as the end-state when a changed organization was accepted and supported by the external environment.[38] But regardless of the intended goal, institution-building projects in practice were largely unsuccessful; often the recipient country did not have the necessary human resource skills and training to maintain projects effectively. High failure rates led to a new emphasis by theorists and practitioners upon capacity building, or the human dimension of political and social organization. Central to the idea of capacity building is a concern for improving indigenous capabilities in order to generate a development impact that surpasses immediate project goals.

Unfortunately, capacity building has been no more successful in practice than institution building. In the words of David Fairman and Michael Ross: 'Perhaps nothing in the field of development is as popular to promote and as difficult to accomplish as capacity building.'[39] Some success has been achieved in relation to financial and technological capacities, but efforts to improve the maintenance of physical infrastructure and interagency coordination remain largely unsuccessful.[40] One of the problems with capacity building has been the contradiction involved in building indigenous capacities on the basis of external resources. An over-dependence upon external consultants, for example, runs the risk of undermining local authority.[41] The problem is particularly acute in relation to international environmental assistance because recipient countries often lack the necessary experience and expertise in environmental management.[42]

But a more fundamental problem with capacity building is that both scholars and practitioners have widely different interpretations of what the concept actually means. Alongside the concepts of 'participation', 'empowerment' and 'ownership', capacity building is seen as essential to development but difficult to define and conceptualize. Current interpretations amongst donor agencies and NGOs range from the very general 'helping people to help themselves' philosophy to the more specific belief in 'strengthening civil society organizations in order to foster democratization, and building strong, effective and accountable institutions of government'.[43] Many commentators now use the terms institution building, capacity building and development interchangeably.

In response to the lack of analytical clarity and precision, John Cohen, amongst others, advocates a return to the traditional minimalist interpretation of capacity building defined as:

> The strengthening of the capability of chief administrative officers, department and agency heads, and program managers in general purpose government to plan, implement, manage or evaluate policies, strategies, or programs designed to [have a positive effect upon] social conditions in the community.[44]

But a simple return to this narrower administrative definition does not solve the problems associated with the concept. This definition assumes that the environment within which administrative officers perform tasks is conducive to change and performance improvements. As argued by Merilee Grindle and Mary Hildebrand, it is impossible to focus upon administrative capacity as an autonomous entity detached from the politics and structural conditions of the local setting as these, in turn, sustain capacity building efforts.[45] Building capacities in order to perform a specific task is not enough; if capacity building is to succeed it must also endure over time.

In taking into account the importance of contextual factors and sustainability, it is useful to think of capacity building as both a means and an end to development.[46] It can have an instrumental role in contributing to the effective implementation of a particular development project, or an intrinsic role in facilitating a broad range of capacities – human, technological, financial or institutional – that are sustainable in the longer term. Instrumental capacity building is project specific and more evanescent. In contrast, intrinsic capacity building focuses upon strengthening capabilities beyond the purpose of the project and therefore has far reaching implications for development per se.

Amartya Sen has been influential in propounding the notion that a 'capabilities approach' is synonymous with development.[47] He argues that the purpose of development is to improve people's quality of life, which is best achieved by providing them with stronger capabilities. His view has been echoed by the OECD which in 1995 pledged to place capacity development at the forefront of development policy.[48] The problem is that many development organizations are now updating their terminology from capacity building to capacity development while leaving the concept intact.

In relation to the environment, the notion of intrinsic capacity building is especially important because managing environmental problems requires a long-term commitment towards building political consensus, institutions and public awareness. The issue of endurance over time is critical. Given that the idea of *environmental capacity building* has been largely neglected in the literature, an initial discussion of the concept can be based on a broad understanding of the difficulties involved in managing environmental problems, especially in developing countries.[49]

At a minimum, weak environmental capacities are rooted in a lack of financial and technological resources, weak environmental bureaucracies, inadequate knowledge and information flows, and the opposition of entrenched political and societal interests. Environmental capacity building is thus an integrative concept that can be broadly defined as *improving the management of environmental problems by strengthening the prerequisite financial, technological and human capacities so that they can endure over time*. To construct a framework for understanding what constitutes capacity building for environmental purposes, the analyst needs to identify the key dimensions in a more disaggregated and systemic way. In other words, we need to know what capacities are necessary.

The key dimensions of environmental capacity

What then constitutes environmental capacity? Given that the environment is still a relatively new field for developing countries, this is a difficult question to address. Currently environmental policies and action plans are mostly being carried out on a trial and error basis and empirical research is limited.[50] But by drawing upon a wide range of scholarship in development theory, international relations and comparative politics, it is possible to distinguish four key dimensions.

The first dimension concerns *financial efficiency*, comprising the mobilization and utilization of funding. The scarcity of financial resources for environmental management is the most obvious capacity constraint in developing countries and one that captures the most attention at international environmental conventions. The issue of the utilization of funding attracts less attention but is equally, if not more, important. One point to stress here is that if efforts to build financial capacity are to be maintained, they must include requirements for long-term recurrent financing.[51] The need for recurrent expenditure is a common theme in the development literature.[52] Yet, in practice, this is often not possible either because of a shortage of funds on the part of the external donor, or because of fiscal constraints within the recipient country. When the tax base of a developing country is weak or politically volatile a secure means of ensuring operational and maintenance funding that does not rely upon external funding is to increase prices.

A second dimension of capacity requirements concerns *institutional cohesion*. The institutional support for environmental management involves strengthening interagency coordination and ensuring greater openness through the exchange of information and participatory-style consultation. International relations scholars Ronnie Lipschutz and Ken Conca argue that the state's weak capacity to deal with environmental problems is based upon its weak institutions.[53] This perception is also a major driving force behind the UNDP's Capacity 21 Fund and the World Bank's efforts to strengthen environmental management in developing countries.[54] Institutional development is considered to be particularly important in the former communist countries of Central and Eastern Europe because of the legacy of top-down policy-making and the lack of experience in creating partnerships among different players.[55]

It is currently fashionable to focus upon institutional development but this cannot be disentangled from political power: neither institutional innovations nor highly skilled environmental professionals are likely to withstand the negative effects of special interest groups and corruption if the political will of the government does not protect them.[56] Consequently, the institutional dimension of environmental capacity building must take political commitment as given. This is not to suggest that international environmental aid cannot have a positive effect upon political commitment. In the situation where the recipient government favours reforms but is constrained by its domestic opponents, evidence suggests that external environmental funds could enhance political capacity. 'In these settings, funders can take the blame for unpopular measures, provide money to buy out opponents to reform, and wield their influence to build environmental alliances.'[57] In other words, domestic proponents can use external funding and advice as a means of boosting their political capital and thereby alter the nature of coalitional politics. However, in such cases, it is important to recognize that building political capacity is the prerequisite for institutional development rather than an end in itself.

A third dimension is *technological advance*. The development literature provides numerous examples of weak technological capacity in relation to the environment, which will not be repeated here. The proponents of the conservative version of the theory of *ecological modernization* discussed earlier, draw attention to many of the benefits that can be gained from improvements in industrial processes and products. Technological innovation is considered by many to be the most cost-effective way of enhancing environmental management in developing countries, particularly in relation to sustainable energy use.[58] Building technological capacity requires knowledge creation and acquisition, which is relatively easy to facilitate through improved monitoring procedures and technical auditing expertise. The diffusion of technological capacity is more difficult. As yet, aside from relying upon the elusive 'demonstration effect', the most appropriate means of facilitating technological diffusion has not been clearly identified.

A fourth dimension refers to *information sharing* and the ways in which environmental knowledge is acquired, interpreted and disseminated. Many theorists have argued that environmental knowledge and information flows are an indispensable means of redefining political interests and enhancing public awareness. Empirical studies on the relationship between knowledge and politics in the environmental domain have largely centred upon the notion of epistemic communities.[59] More recently, the notion of disclosure strategies (social incentives) involving public and/or private attempts to increase information on pollution has formed the so-called third wave in pollution control policy thinking.[60] The 'right to know' emphasis is being promoted in a number of OECD and developing countries in recognition of the fact that information has proven to be instrumental in improving public participation in regulatory enforcement.[61] In this respect, Martin Jänicke and Hans Weidner note that the role of the media is important as both a provider of information and as a catalyst for public environmental awareness.[62]

I would argue, however, that too much emphasis upon information sharing distracts attention away from the importance of social norms. Better information may increase environmental awareness but it does not necessarily lead to changes in behaviour. For this we need a better understanding of the social and normative context in which information exchange takes place. Building informational capacity is important but it should not be taken as a substitute for the more difficult task of building environmental constituencies and nurturing stakeholder participation. To be effective over time, improving the exchange of information should go hand in hand with nurturing a sense of responsibility and, therefore, a desire to take action on the part of the stakeholders involved. Hence the informational requirement of environmental capacity needs to be subsumed under a broader social dimension.

These four dimensions of environmental capacity – financial, institutional, technological and social – are not exclusive domains but are mutually reinforcing. Efforts to build financial capacity, for example, are likely to have consequences for institutional capacity and vice versa. At this stage, the framework needs refining on the basis of empirical analysis. The social dimension, in particular, requires further clarification and development. Empirical investigations are also needed to identify how environmental capacity building can best be measured. In practice, how do we know when environmental capacities are actually being developed? To date, donors have tended to either ignore the issue or focus on a narrow set of indicators related to human resource development. For the purposes of this study, a number of broad indicators will be used such as actual improvements in pricing structures, interagency coordination, the cost-effective use of technologies or monitoring activities. Local perceptions and changes in attitude will also be an important means of assessing the contribution of different donor approaches to building more sustainable environmental capacity.

Conclusion

This chapter has identified two major limitations with the theoretical discourse on environmental management. First, it places too much emphasis upon policy principles while largely ignoring the problems of implementation in diverse political and socio-economic settings. While the market analytical construct differs from the participatory approach over the question of how the environment should be managed, they both advocate less intervention from the state. But the reality is that in both developed and developing countries the state plays an essential role in environmental management. Indeed, despite contentious debate between economists, political scientists and environmentalists over the problems involved in central planning, in practice state-led regulation still remains the dominant mode of managing environmental problems globally.

It is the argument of this book that a pluralistic approach to environmental management is more likely to generate effective environmental outcomes. This is because a market-oriented or participatory approach in isolation cannot guarantee behavioural change. Moreover, political and economic power constraints are more likely to be resolved by combining state regulation with creative economic incentive schemes, participatory practices and innovative technological solutions.

A second limitation with the theoretical discourse is that it fails to explain how policy principles relate to capacity building. The conceptual framework proposed in this chapter is an initial attempt to address this caveat. It will be used to interpret the effects of various donor environmental approaches on local capacity in China, and will then be further refined in Chapter 6 on the basis of the empirical investigations. Let us now take a closer look at China's evolving state-centric system of environmental management, and the potential for building capacity at the local level.

2 The long march towards environmental management in China

Environmental management in China is dominated by state regulatory control. It is also biased towards dealing with the effects rather than the causes of environmental degradation. China's historic legacy of large-scale engineering projects casts a looming shadow over its environmental planning and investment agenda. Nevertheless, this chapter will show that the Chinese government is moving towards a broader interpretation of environmental management that embraces regulations, legislation, economic incentives and, to a lesser degree, public participation. In effect, at least at the policy level, environmental reforms in China are increasingly aligned to international standards.

However, both market incentives and public participation remain difficult options in practice. The prospects for successfully implementing a market-oriented approach to environmental management are higher in China's richer regions where economic reforms are more advanced and where the ability to pay for environmental quality is relatively high. The prospects for implementing a participatory approach are less clear. Although a green civil society is emerging in China, citizen involvement in environmental management remains heavily circumscribed.

A more immediate constraint seems to be that a disjuncture exists between ongoing environmental reform at the central level and poor implementation at the local level. Conventional wisdom dictates that protecting the environment is more effective if it is relevant to local circumstances. But in China, economic decentralization has not provided the positive environmental benefits that one might expect. Instead, the economic reform process has had a contradictory impact on environmental reform: fiscal and administrative decentralization has provided local officials with the authority and resources to manage environmental problems, and, at the same time, has empowered entrenched local and political interests against the imposition of state regulatory controls.[1]

The one issue that the majority of China scholars find it easy to agree on is that environmental improvement in China is severely constrained by weak compliance at the local level. This chapter will argue against the prevailing scepticism by questioning the assumption that weak compliance is the only impediment to effective environmental management. In so doing, it will provide some insights to suggest that weak capacity could also be a major constraint. The chapter begins by looking at the severity of environmental problems in China followed by an in-depth discussion of the nature of China's evolving state-centric system of environmental management. The central focus of inquiry then shifts to the difficulties involved in managing environmental problems at the local level. It concludes by considering two important implications for the delivery of international aid.

The scale and severity of environmental problems in China

In recent years, environmental issues in China have captured the attention of a growing number of Western and Chinese scholars. Some of the earlier works in the 1970s were extremely positive about environmental management, even seeking to present Chinese traditional and socialist practices as a model for the world.[2] In the 1980s and early 1990s, as more reliable information became available, the magnitude of China's environmental problems slowly came to light.[3] More recently, official reports and the Chinese media have played an important role in revealing the severity of the crisis.

The following survey presents a brief historic overview of environmental problems in China and summarizes some of the key current trends. In keeping with the focus of this study, special attention is given to pollution concerns. The discussion is based on official statistics from the State Statistical Bureau (SSB) and the State Environmental Protection Administration (SEPA). The validity of Chinese environmental statistics, however, remains questionable. Although the reliability of pollution monitoring has increased considerably over the past decade, and is more in keeping with international standards, caveats remain.[4] There is a lack of data on pollution emissions from township and village enterprises (TVEs), rural and semi-urban pollution in general, and on losses in biodiversity.

Historic environmental legacies

As many commentators have observed, environmental degradation in China is not new; the environment did not drastically alter at the onset of rapid economic growth in the early 1980s. Instead, environmental problems have their origins in China's history. Since ancient times, frequent flooding in China has encouraged the development of innovative hydraulic engineering practices.[5] Concerns over fuel supplies also encouraged large-scale reforestation. As early as the Qin (221–206 BC) and Han period (206 BC–AD 220), shelter belts (tree plantations) were planted alongside the Great Wall by Emperor Qin Shihuang, and by Qing times (1645–1911) widespread reforestation was taking place.[6]

Water and forestry conservation, however, did not prevent environmental mismanagement. Over the centuries, China's ecological balance became disrupted by the extraction of natural resources in the pursuit of state military power and excessive land cultivation.[7] By the nineteenth century, China's environmental history is a familiar story of converting forests and wetlands to fields leading to increased siltation in rivers and, in turn, more flooding. One well-known environmental historian claims China avoided ecological breakdown only through 'spatial expansion and technical breakthrough'.[8] By the time the Communists came to power in 1949 the seeds of China's current environmental crisis had already been sown.

Initially, shelter belt planting to protect against soil erosion continued into the 1950s, culminating in the Three North Shelter project (*san bei fang lin*). But the Great Leap Forward (1958–1961) and the Cultural Revolution decade (1966–1976) wreaked havoc upon the environment and more than outweighed conservation efforts. According to Shinkichi Eto, 13 million hectares of forests were cleared during the Great Leap Forward to fuel backyard iron smelters and only 240,000 hectares of forests were replanted; the famine that ensued led to over 30 million deaths – the worst recorded in history.[9]

Before the proclamation of the People's Republic of China, Mao Zedong wrote, 'It is a good thing that China has a big population. Even if China's population multiplies many times, she is fully capable of finding a solution; the solution is production.'[10] Production became the *zeitgeist* of the Chinese Communist Party (CCP) and China's population increased from 574 million in 1952 to 829 million in 1970. Heavy industrialization was the overriding priority. The Party implemented draconian procurement policies to keep agricultural prices artificially low in order to develop the cities and industry.[11] Furthermore, in the spirit of '*xian shengchan, hou shenghuo*' (putting production first and the standard of living second) urban services and infrastructure were severely neglected. To this day, access to safe water and sanitation is amongst the worst in the world.[12]

Despite the obvious negative environmental effects of Mao's policies,[13] political rhetoric at the time defined environmental pollution as a purely capitalist construct: 'Capitalists…discharge at will and in disregard of the fate of the people, harmful substances that pollute and poison the environment'.[14] Western commentators such as William Kapp supported such a distorted view by praising the Chinese concept of development, which allegedly prioritized social welfare above technical advancement.[15] In reality, under socialism – as with capitalism – natural resources were treated as capital rather than income; production targets mattered and not the net social benefits.

The historic legacies of China's environmental mismanagement remain. First, the Chinese government is still strongly preoccupied with water control and conservation which, in turn, reinforces an engineering mindset in regard to environmental protection. Second, China still suffers from a lack of environmental infrastructure, which necessitates an investment priority for end-of-pipe environmental solutions as opposed to capacity building. Finally, Stalinist-style industrial behemoths are still major polluters. Since the onset of the economic reforms in the early 1980s, China has suffered the worst of both worlds, caught between the inefficiency of state-owned enterprises and thousands of uncontrollable capitalist TVEs.[16]

Current environmental trends

Although China has a vast territory (9.6 million km^2) with huge natural resources, its per capita resource base is low – 22 per cent of the world's population lives on

7 per cent of the world's arable land with only 8 per cent of the world's natural resources.[17] China's natural resources are also unevenly distributed: the areas south of the Yangzte River have 80 per cent of the nation's water resources but lack land, while the reverse situation exists in the north of China.[18] Consequently, 27 per cent of China's grain is produced in the north leading to the over-exploitation of ground water resources and a dramatic reduction in the water table – the north faces serious water shortages while the south experiences frequent flooding.[19]

The environmental effects of China's population pressure combined with unsustainable practices are palpable.[20] According to SEPA's annual report released in June 1999, 90 per cent of China's grasslands have become degraded and desertification now covers one-third of China's land base. In addition, almost 40 per cent of the nation suffers from soil erosion as unsustainable mining and farming practices cause 10,000 km^2 of lost soil per annum. Forest resources are scarce (14 per cent coverage, representing 4 per cent of the world's total) and wetlands have been reduced by 60 per cent. Excessive land cultivation (particularly in semi-arid sloping lands) and shortages of fuel wood are the main causes of deforestation. Grasslands are under threat because farmers lack sufficient incentives to maintain sustainable grazing levels, while wetlands in China are still perceived as wastelands rather than sinks for pollutants, buffers from storms or fish nurseries. The loss of biodiversity in China is equally serious. Although exact figures are not available, it is estimated that of the 640 species listed under the United Nations Convention on International Trade in Endangered Species of Wild Fauna and Flora (CITES)[21] approximately 156 are in China.[22]

The scale and severity of China's pollution problems are equally sobering. Approximately half of the monitored urban river sections in northern China (Huang, Huai, Hai and Liao) do not meet China's lowest ambient standard (grade 5) – the water is officially classified as unfit even for irrigation. In addition, total suspended particulates (TSP)[23] or airborne particulates and SO$_2$ emissions are amongst the highest in the world – in many cities exceeding World Health Organization (WHO) recommendations by 2–5 times (see Table 2.1). In some regions the levels of TSP are dangerously high, especially in China's northern cities. In 1997, TSP levels in Taiyuan, Shenyang and Lanzhou were approximately 600 μg/m^3 – ten times the international standard.

The health effects are devastating. Nearly 700 million Chinese drink contaminated water; and an estimated 178,000 people in China's 560 major cities suffer premature deaths each year due to air pollution.[24] SO$_2$ pollution can lead to increased incidence of bronchitis, respiratory diseases and emphysema. It is also a precursor of acid rain, which causes the corrosion of metals and chronic plant deterioration. Acid rain now covers roughly 30 per cent of land area in China. It is a particular problem in the south-central and southeastern regions of China because of naturally acidic soils and the high sulphur content of coal reserves (equal to 4 per cent). Consequently, the average pH value (below 5 = acid rain) of precipitation is often 4 or even 3 (equivalent to vinegar). The lowest recorded values are in Liuzhou in Guangxi Zhuang Autonomous Region (3.06) and Guiyang in Guizhou province (3.15).

Table 2.1 Chinese ambient air quality standards ($\mu g/m^3$)

Pollutant	Time span	Chinese maximum concentration			WHO standards[b]	National average (1997)
		Class I[a]	Class II	Class III		
Total suspended particulates	Daily average	150	300	500	60–90	309
SO_2	Daily average	20	60	100	40–60	79
NO_x	Daily average	50	100	150	150	46

Source: Adapted from Richard Louis Edmonds, *Patterns of China's Lost Harmony: A Survey of the Country's Environmental Degradation and Protection*, London: Routledge, 1994; and World Bank, *Clear Water and Blue Skies: China's Environment in the Twenty-First Century*, Washington, DC: World Bank, 1997.

Notes
a Class I includes tourist, historic and conservation areas. Class II are residential urban and rural areas. Class III are industrial and heavy traffic areas.
b The World Health Organization Standards are not exactly comparable to Chinese standards as a result of differing methods of monitoring as well as time spans. In the 1990s WHO eliminated its TSP standard because no threshold exists below which health impacts are negligible.

China's urban pollution problems are reminiscent of London in the 1950s or Tokyo in the 1960s, only more serious. In 1997, air pollution in Chongqing was significantly worse than in London in 1952 and SO_2 concentrations in Lanzhou and Chongqing were approximately two and a half times higher than levels recorded in Japan in its peak year.[25] Moreover, levels of organic mercury in the Songhua River in northeast China have been reported to be five times higher than in the Minamata Bay in the 1960s.[26] Typically, Chinese industries are highly energy intensive, heavily polluting and technologically backward. In particular, the chemical industry ranks first amongst all industries in the discharge of toxic substances (such as mercury, phenol, chromium and cyanide (Cn)). Rural pollution is more difficult to monitor but China's growing TVEs are also heavily polluting, and the intensive use of nitrogen fertilizers and pesticides exacerbates the problem. Indoor pollution is consistently high because roughly 80 per cent of rural households still use solid fuels such as coal, firewood and crop stalks. The World Bank estimates that up to 700,000 premature deaths a year could be prevented if households were to switch to using coal briquettes and gaseous fuels in addition to improving ventilation.[27]

More recently some important achievements have been made in both water and air pollution abatement (see Table 2.2). For example, the treatment of industrial wastewater has increased from 72 per cent in 1993 to 95 per cent in 2000 (although less attention has been given to controlling municipal wastewater that now represents 40 per cent of the aggregate total).[28] According to a report by the National Resources Defense Council in New York, CO_2 emissions have declined

Table 2.2 Key indicators of industrial pollution abatement in China[a]

Year	%
Share of coal consumption in total energy consumption	
1957	92.3
1985	75.8
1993	72.8
1996	75.5
2000	69.0
Industrial wastewater treatment ratio[b]	
1981	13.0
1985	22.1
1993	72.0
1994	75.0
1995	76.8
1996	81.6
2000	95.0
Industrial solid waste utilization ratio	
1985	33.0
1990	35.0
1993	38.7
1995	43.0
1996	43.0
2000	51.8
Industrial waste gas treatment ratio	
1985	20.2
1995	70.8
1996	75.0
2000	82.6

Source: Compiled by author from *Zhongguo tongji nianjian* 1997, 2000 (China Statistical Yearbook), Beijing: Zhongguo tongji chubanshe; *Zhongguo huanjing nianjian* 1996, 1997, 2000 (China Environmental Yearbooks), Beijing: Zhongguo huanjing chubanshe; and Alice H. Amsden, Dongyi Liu and Xiaoming Zhang, 'China's macroeconomy, environment, and alternative transition model', *World Development*, 24(2), 1996.

Notes
a Statistics are based on data collected from industrial enterprises at and above the county level.
b Note here that the triple increase in treatment ratios between 1985 and 1995 can be accounted for by the implementation of environmental standards. The percentage of industrial wastes actually meeting those standards, however, is lower than the level of treatment. For example, in 1995 the percentage of wastewater treated up to standard was 58.6 per cent compared with 76.8 per cent treated in total.

Table 2.3 Total emissions of TSP and SO$_2$ in
China (millions of metric tonnes)

Year	TSP	SO$_2$
1988	14.60	15.20
1992	14.10	16.90
1996	13.35	13.63
1999	10.15	10.78
2000	9.21	11.49

Source: Compiled by author from *Zhongguo tongji
nianjian*, various years (China Statistical Yearbook),
Beijing: Zhongguo tongji chubanshe; Z.X. Zhang and
Henk Folmer, 'The Chinese energy system: implica-
tions for future carbon dioxide emissions in China',
Journal of Energy and Development, 21(1), 1996; and
China Daily, 4 June 1998.

dramatically by approximately 17 per cent since 1997.[29] TSP levels and SO$_2$
emissions have also decreased significantly since they peaked in the mid-1990s
(see Table 2.3). According to SEPA, between 1998 and 2000 SO$_2$ emissions
declined by 1.86 million tons and discharges of 12 other major pollutants
decreased by 15 per cent.[30] Overall emissions reductions are largely a result of
fuel switching, increasing residential use of natural gas, expanded district heating,
a lower level of coal consumption[31] and regulatory control.[32]

On a less positive note, nitrogen oxide (NO$_x$) emissions are rising dramatically
in the major cities of Shanghai, Beijing and Guangzhou (between 2000 and 2001
the number of vehicles in Beijing alone increased by 82,000 to 1.8 million).[33] As
noted by Michael McElroy *et al.*, these cities are now experiencing Los Angeles
style photochemical smog which is highly toxic and can lead to serious health
problems.[34] A study carried out by Xiao-ming Shen *et al.* in 1996 provided an
early warning sign of the health effects arising from lead poisoning. The study
revealed that children in Shenyang, Shanghai, Guangzhou and other major cities
had blood-lead levels averaging 80 per cent higher than levels considered dangerous
to mental health by WHO.[35]

Estimates of the economic costs of China's environmental degradation have
varied over time according to the perceived severity of the threat and the methodologies employed. In 1993, a report by the Chinese Academy of Social Sciences
estimated that pollution cost, RMB102.92 billion (US$12.5 billion)[36] per annum
in direct losses.[37] According to the World Bank, air and water pollution damages
to human health alone equalled almost 8 per cent of Gross Domestic Product
(GDP) in 1995. Chinese estimates were significantly lower at 3.5 per cent of GDP.[38]
A year later, Vaclav Smil, one of the leading foreign experts on China's environment, estimated the real costs of environmental degradation to be in the range of
10–15 per cent of gross domestic product (GDP).[39] By 1999, official Chinese
environmental accounting appeared to be more realistic. The economic losses

from SO_2 emissions and acid rain were officially estimated to total more than RMB110 billion (US$13 billion) per annum.[40] In addition, Chinese officials claimed that the losses from natural disasters cost at least 3 per cent of GDP per annum.[41] The flooding of the Yangtze River in 1998 led to an official death toll of 3,000 with five million homes destroyed and 52 million acres of land inundated. The estimated economic cost was around US$20 billion.[42]

Without doubt, China is experiencing considerable environmental stress. Despite some achievements to date – notably the increase in industrial waste treatment, higher energy efficiency, the shift towards the utilization of household gas and reductions in emissions – pollution trends are still extremely serious by international standards and ecological degradation remains unchecked. This raises an important question: to what extent does China's environmental crisis threaten the economic and social foundations of its socialist modernization? If current environmental trends persist, it is unlikely that China will be able to sustain its rapid development trajectory. However, future scenarios are difficult to predict without an in-depth understanding of both the willingness and capacity of the Chinese government to implement environmental reforms. Let us now look at how the Chinese government has responded to its environmental crisis before discussing the problems of implementation at the local level.

Managing the environment from above

Until the mid-1990s, China's concern with environmental issues had been partial at best; far greater emphasis had been placed upon economic imperatives in the Deng Xiaoping spirit of 'to get rich is glorious'. The Chinese government's resistance to environmental protection was predicated on the assumption that rising incomes would inevitably lead to rising environmental concern – 'the more rich, the more ecological' thesis. For example, the Chinese spokesperson for environmental affairs, Dr Song Jian, said at the UNCED in 1992, 'one should not give up eating for fear of choking'.[43]

Despite this apparent ambivalence, in 1993 environmental protection became a fundamental state policy (*jiben guoce*). Concern for the environment became particularly pronounced in 1996, signalled by the first appearance of a Chinese President, Jiang Zemin, at the Fourth National Conference on Environmental Protection in July 1996. An extract from the General Secretary's speech contained the following:

> Some comrades ignore the work of environmental protection. They believe that it is the most important to develop the economy and environmental protection can be put aside at the moment . . . many economically developed countries took the same approach of serious waste of resources and 'treatment after pollution', thus causing grave damage to the resources and ecological environment of the world. This is a serious lesson drawn from the developed world. We should never take such an approach.[44]

Rhetoric in this case has been matched by practice: investment in environmental protection as a percentage of GNP rose from 0.67 per cent during the seventh Five-Year Plan (1986–1990) to a pledged 1.5 per cent during the tenth Five-Year Plan (2001–2005). This is equal to the current average investment ratios of the industrialized OECD countries. In 1996, environmental priorities were also integrated into the state's Ninth Five-Year and Long Term Plans to 2010. Then in 1998, state environmental concerns further increased in light of four environmental disasters: the appearance of red tides in the Bohai Sea, severe flooding in the Yangtze, Songhua and Nen river basins, foam floating on the *Huang He* (Yellow River), and sandstorms in Inner Mongolia and Xinjiang. In response to the 1998 floods, the State Council endorsed a National Plan for Ecological Environmental Construction forbidding logging in natural forests and the reclamation of land for farming. This represented a slight shift in policy thinking away from pollution control and towards natural resource management.

The motives behind environmental reform

The dilemma facing the Chinese leadership is that investing in the environment is costly and could reduce economic competitiveness, but in the longer term, a failure to address environmental degradation could lead to serious economic decline, social tension and, in turn, the loss of political legitimacy. A greater understanding of international practices seems to have increased government awareness that prevention is better than cure (i.e. the costs of industrial retrofitting far outweigh the costs of investment in environmental technologies, and likewise effective watershed management protects reservoirs from siltation and floods and is a far cheaper option than the rehabilitation of a deforested, eroded watershed).

The opportunity to procure international financial and technological investment provides an additional economic incentive, although this should not be exaggerated. The scale and severity of environmental problems in China mean that domestic responsibility is imperative. In the words of Song Jian, then Chairman of the State Council's Environmental Committee, at a press conference in 1991: 'while we are appealing to the developed countries for help in environmental protection, we will depend mainly on ourselves'.[45] In some ways this is not surprising given that environmental protection is consistent with the idea of a Chinese socialist market economy and the need to shift from extensive to intensive modes of production.

Beyond the economic incentive lies a strong political rationale that appears to have tipped the balance more firmly in favour of environmental reform. The Chinese leadership has not been slow to advertise its green credentials as a means of winning domestic approval. Arguably, the environment has provided a new form of political legitimacy: China's mass organizations (*qunzhoung tuanti*) now have a new issue to promote as a means of rallying public support; the People's Liberation Army and party organizations at all levels are equally promoting the environmental cause as a means of purchasing legitimacy.[46]

National pride is also a strong motivating factor. Environmental problems are difficult to ignore because of their visibility.[47] Both the positive and negative effects of China's rapid economic growth over the past 25 years are now on show to the outside world. China's promise to clean up its environment was a major factor in winning approval to host the Olympic Games in Beijing in 2008. Arguably, this has provided a far greater incentive than any amount of international environmental funds. Recent plans under the Green Olympics Campaign include transmitting natural gas from northern Shanxi to households in Beijing, substituting oil for liquid petroleum gas (LPG) on public transport, banning automobiles that fail exhaust emission standards, constructing a fourth ring road to divert traffic from the centre, and relocating or closing down factories. The Beijing municipal government plans to spend a total of US$12.2 billion on clean up costs by 2008.[48] The Campaign, however, does have its environmental downside. A number of unofficial reports by Chinese environmental NGOs suggest that the planned face-lift for Beijing masks serious problems in integrated and sustainable urban planning.

In sum, China's rising concern for the environment at the central level has been motivated by a mix of political, economic and psychological factors. The central importance of international influence has been in providing the Chinese leadership with an understanding of the economic costs of not integrating environmental concerns into its development policy-making, and by demonstrating how (in theory) China can continue to pursue economic growth and environmental protection simultaneously.[49] International influences have also had some effect on environmental policy-making in China, especially in the areas of environmental legislation and market incentives.[50] China's incremental 'trial and error' approach to the environment has led to an eclectic policy-making system that is essentially a mix of engineering solutions, regulations, legislation and new non-state practices. Let us first look at the fundamentals of environmental management in China before turning to its more innovative dimensions.

The engineering mindset

The ongoing construction of the Three Gorges Dam (*San Xia*) – 200 m high and 2 km wide – in the middle reaches of the Yangtze River is symbolic of China's historic legacy of environmental engineering. Total construction costs are predicted to be in the range of RMB95–120 billion (US$11.5–$14.5 billion), with a construction period of approximately 17 years.[51] The dam is expected to contribute 3 per cent to China's primary energy demand and 7 per cent to electricity production. The human and ecological implications are vast: officially, 1.2 million people will need to be relocated, with critics suggesting that the real figure will be more in the range of 1.5–2 million.[52] The Yangtze Dolphin, an endangered species, will be further threatened and much of China's finest scenery and wetlands habitat destroyed. More worrying is the fact that silt build-up behind the dam might place it under jeopardy, blocking downstream regions of vital nutrients as well as stalling power generation.[53]

Less well known is the south–north water transfer scheme (*nanshui beidiao*), which plans to divert water from the Yangtze River to the North China Plain. Three planned routes have been designed to connect the Yangtze with the Huang, Huai and Hai rivers. The construction of each route will be carried out in three phases at a total estimated cost of RMB180 billion (US$22 billion).[54] The project aims to alleviate flooding in the south and water scarcity in the north but, again, the ecological consequences are potentially severe. Possible problems include large-scale soil salinization, salt water or polluted sewage water intrusion and adverse effects upon aquatic life in ponds and lakes along the route.[55]

This emphasis upon an 'engineering fix' is hardly surprising given the backgrounds of the key members within the Chinese leadership. As noted by Lyman Miller, in 2002, 18 out of the 24 Politburo members were 'engineers by training'.[56] Government officials involved in environmental planning in China are also mostly engineers – usually trained in the former Soviet Union. On the basis of this distinctive engineering outlook, the recent investment drive in environmental protection in China centres mostly upon infrastructure building and traditional end-of-pipe solutions.

In 1996, SEPA launched China's TransCentury Green Plan (1996–2010)[57] that in its first phase (1996–2000) focused upon controlling pollution – largely by investing in engineering projects – in three lakes (Taihu, Chaohu and Dianchi), three rivers (Huai, Hai and Liao), and two control regions for acid rain and SO_2. Planned investment for 1996–2000 totalled RMB4500 billion (US$54 billion at 1995 prices – equal to 1.3 per cent of GDP). This included RMB200 billion targeted for new pollution projects or renovation, RMB145 billion for urban environmental infrastructure, and RMB105 billion for old pollution sources. Investment targeted for capacity building (defined as strengthening county level environmental protection bureaux and increasing monitoring stations) was considerably lower at RMB3.25 billion. Actual environmental investment during this period reached approximately 0.9 per cent of GDP in 2000.[58]

Planned environmental investment under the tenth Five-Year Plan is still heavily focused on pollution control and engineering projects, especially in relation to improving water pollution in large river basins and urban air quality management. Other priorities include the control of non-point pollution sources in China's rural areas and combating land degradation. As noted by the World Bank, projected investment in ecological conservation is expected to account for only 7 per cent of total environmental expenditure. Planned investment in capacity building is even lower at 1.5 per cent.[59] Of course, this strong engineering focus could easily be justified on the grounds of China's huge environmental infrastructure deficit but, nevertheless, it is difficult to see how it can make the transition away from environmental engineering and towards comprehensive environmental management without the necessary investment in building institutional and managerial capacities.

State regulatory control

In March 1998, China's environmental bureaucracy was reformed as part of the administrative restructuring process initiated at the Ninth National People's Congress.[60] The Environmental Protection and Resources Committee of the National People's Congress (China's legislature) continues to play a key role in drafting and revising environmental legislation, but the State Environmental

Protection Commission that was responsible for interagency policy coordination was disbanded. Coordinating responsibility now rests with SEPA, which in turn has been upgraded to ministerial status.

As the key administrative authority for environmental protection in China, SEPA is responsible for nationwide supervision and administration. It provides vertically integrated support for local environmental protection bureaux (EPBs) at the provincial, municipal, county and township levels (see Figure 2.1). As part of the reforms, its functional role expanded from predominantly pollution control to include nature reserve management, biodiversity, desertification, wetland conservation, mining, marine pollution and nuclear safety. Given that the organization

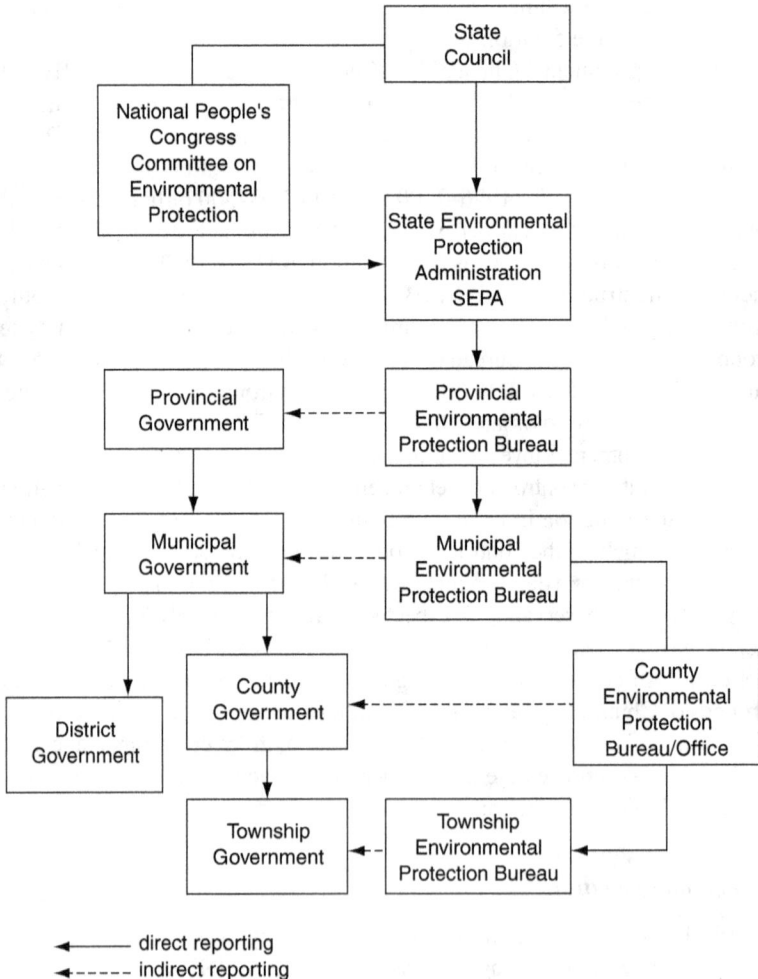

Figure 2.1 China's environmental regulatory framework.

Source: Author.

has had to face a 37 per cent reduction in its staff,[61] the ability to carry out these responsibilities, however, remains under question.

Environmental protection offices have also been established within key state agencies, such as the National Development and Reform Commission (NDRC), the State Economic and Trade Commission (SETC), and the Ministry of Science and Technology (MST).[62] Sustainable development is promoted by NDRC and MST at the central level, and mostly by the planning commissions at the local level.

Environmental policy-making in China is based upon the dual principles of 'prevention first' and 'polluter pays'. The environmental impact assessment (EIA) methodology was adopted from developed countries and is a mandatory policy for all medium and large investment projects. Other key policy instruments include the 'three simultaneities' (*san tongshi*), pollution fees and discharge permits (see Table 2.4).[63] The *san tongshi* policy requires pollution controls to be integrated into the design, construction and operation phases of all investment projects. Pollution fees are charges levied on enterprises for emissions exceeding national standards – 80 per cent of fees are returned to the enterprise on the condition that the funding is used exclusively for environmental protection. The remaining 20 per cent of fees are retained by the EPBs.

Table 2.4 China's environmental policies

1979–1989	1989–1996	1996–2002
• Effluents and emissions standards • Environmental impact assessment • *San tong shi* (pollution control facilities to be included in the design, construction and operation phases of all new construction projects) • Discharge fees (pollution fees set at below the marginal costs of controlling emissions and not index-linked – 80 per cent refunded back into firms for investment in environmental protection)	• Discharge permit system (fines for failing to meet targets based upon the concentration of pollutants in the local area) • Environmental units established at enterprise level to monitor emissions • Environmental responsibility system (rewards/punishments for local leaders and government officials meeting environmental targets)	• Discharge permit system targets total pollution loads • Three Green Project (promotion of a market for environmentally sound produce, environmental labelling and better quality food) • Tradable permit system (on trial) • SO_2 quota system (on trial) • Cleaner production • Pollution emission fee management system (fees collected on the basis of the environmental impact and used to compensate environmental damage)

Source: Compiled from Theodore Panayotou, *Instruments of Change: Motivating and Financing Sustainable Development*, London: Earthscan, 1998; Eduard B. Vermeer, 'Industrial pollution in China and remedial polices', *China Quarterly*, 156(December) 1998; and *Zhongguo huanjing bao* (China Environment News), various editions.

Discharge permits target the concentration of pollutants, or more recently total pollution loads. A nationwide system for wastewater discharge permits was established in 1995 and the enforcement of SO_2 discharge fees began in 1993.[64] Funds released under the environmental discharge system are reported to account for 16 per cent of the total cost of pollution in China – with 25–35 per cent in some cities. But regional implementation varies significantly. According to a report carried out by the State Auditing Administration in 2000, 16 out of 46 key cities in China were believed to have misappropriated funds collected for environmental purposes.[65]

Outside of China's major cities regulatory control is even more problematic, especially in relation to the burgeoning private/collectively owned sector or TVEs (*xiangzhen qiye*).[66] TVEs provide a crucial means of improving the livelihoods of rural Chinese thus preventing large-scale migration to the cities; they absorb labour and provide local income through taxation; and represent the engine of growth behind China's export drive.[67] But these enterprises are also heavily polluting – typically involved in resource intensive industries such as pulp and paper, food processing, printing, dyeing and electroplating. The government's response has been to close down some of the worst offenders, although it would seem that, to a certain extent, such action has been carried out by default. In 1996, the State Council ordered the closure of small paper factories (below 5,000 tons output per annum), leather factories (output below 30,000 hides per annum) and dyeing factories (below 500 tons per annum) as well as chemical and electroplating industries.[68] Clearly these enterprises fall as much into the category of 'non-performing enterprises' in need of closure as they do into the category of 'heavily polluting enterprises'. By 1997, over 64,000 TVEs had been closed down.[69]

However, Eduard Vermeer has made the important point that closing down factories does not necessarily improve environmental quality because of the risk of displacement.[70] The policy has also been contested by some environmental economists in Beijing; they reason that polluting factories would prefer to pay taxes rather than be closed down, thus providing an opportunity for China to implement environmental taxes.[71] Overall, the closure policy is seen to be socially disruptive and ineffective on a long-term basis.

Environmental legislation

In addition to the above policies, the law plays a major role in environmental management in China which, in turn, has reinforced administrative authority. The director of SEPA, Xie Zhenhua, is a major driving force behind the government's keen interest in environmental legislation. He is also a lawyer who strongly believes in the legal axiom *ubi jus, ibi remedium*. In his words:

> [p]ublic investment and participation in the environmental field must be supervised under law. This requires that the public understand the relevant laws and what rights and responsibilities individuals must take to protect the environment.... Our best work and strongest support is a reliance upon a legal practice and supervisory system.[72]

Table 2.5 Major Chinese environmental laws 1979–2002

Year	Law
1979	Environmental Protection Law (Trial)
1981	Marine Environmental Pollution Law
1984	Water Pollution Prevention and Control Law
1985	Grasslands Law
1986	Forestry Law
1987	Air Pollution Law
1988	Water Resources Law
	Law on Protection of Wildlife
1989	Environmental Protection Law
	Noise Pollution Law
1991	Water and Soil Conservation Law
1995	Amended Air Pollution Prevention and Control Law (stronger regulations for controlling SO_2)
	Solid Waste Pollution and Prevention Control Law
1996	Amended Water Pollution Prevention and Control Law (inclusion of river basin pollution and the protection of drinking water resources)
1996	Law on Control and Treatment of Noise Pollution (replacing 1989 regulations)
1997	Flood Control Law
1998	Law on Forests
1998	Wildlife Protection Law
1999	Law on Maritime Environmental Protection
2002	Cleaner Production Law

Source: *Guojia huanbaoju zhengce faguisi* (ed.), *Huanjing baohu fagui huibian* (Compendium of Environmental Protection Laws and Regulations), Beijing: Zhongguo huanjing kexue chubanshe, 1990; and www.enviroinfo.org.cn/LEGIS/Laws/

Since 1979 China has passed more than 40 environmental protection related laws (see Table 2.5), together with a large number of regulations. Particularly noteworthy is the Criminal Law on 'disrupting the protection of environmental resources' adopted at the Eighth National People's Congress in March 1997. Offenders (including government officials in breach of environmental regulations) are now liable for up to seven years imprisonment and substantial fines.[73]

It is common knowledge, however, that China's legal system is inherently weak: legal provisions are often unclear and prone to subjective interpretation, or captured by local interests; and few avenues exist for citizens to seek redress. The majority of legal and regulatory disputes are handled outside of the court system. That said, the Chinese media have reported a small but growing number of cases where citizens have filed law suits against polluting enterprises. For example, in October 2000, fishermen from Laoting county in Hebei took their case to court after losing the majority of their stock when a paper mill upstream in Qian'an county had discharged wastewater into their aquaculture farms.[74] Other court cases have involved a farmer from Huairou County outside of Beijing whose ducks had been poisoned by polluted water from the wastes of a nearby pig farm.[75] Or the orange growers from Maling County in Guizhou who lost their harvest allegedly on account of waste gas from a nearby factory.[76]

Little information exists on the success rate of these cases but the fact that ordinary citizens are pursing a legal means of redress suggests a rise in public confidence in China's legal system. The hope is that these early positive signs will have some kind of cumulative effect and eventually lead to genuine legal reform. The initial momentum is likely to be accelerated on account of China's growing number of legal aid centres. In the case of the duck farmer cited above, a legal aid centre affiliated to the Chinese University of Law and Politics in Beijing provided legal advice and representation in the local court.

At a broader level, appealing to mass consciousness (*qunzhong yishi*) continues to play a major role in state-led environmental management in China, especially in regard to tree planting. The Maoist legacy of 'models' for nationwide emulation still endures – individuals, villages and townships are held up as shining examples for the public to follow. The practice is being further enhanced through international cooperation – a number of villages and townships in Zhejiang province have been awarded Global 500 titles by the United Nations Environment Programme (UNEP) for having 'ecologically friendly environments'.[77]

The problem is that regulatory controls and moral suasion rely upon a state-centric notion of the public good rather than a realistic understanding of the conflict between individual self-interest and social and environmental needs as perceived by the Chinese people. The effectiveness of China's environmental policy-making system, therefore, is likely to depend upon striking a balance between the state, the market and civil society. Let us now turn to a discussion of some of the advances that have been made in implementing non-state approaches to environmental management.

Emerging economic and social dimensions

Since the late 1990s the government has been experimenting with a number of economic incentive schemes. Many of the new schemes are not uniformly implemented. As one environmental economist remarked, 'trading emissions are interpreted differently – many cities simply exchange emissions between old and new plants'.[78] Nevertheless the willingness to experiment is a positive step in the direction of harnessing the market for environmental purposes. In 1999, at least 10 cities in China were trading SO_2 emissions on a trial basis. Market-based taxes (modified to 50 per cent market rate) have now been introduced in the cities of Jilin, Hangzhou and Zhengzhou.[79] And, in early 2002, an SO_2 quota system was set up in seven provinces and cities.[80]

In the spirit of ecological modernization, the government is also actively promoting cleaner production methods in Chinese factories as well as pursuing the development of an indigenous environmental protection industry.[81] These initiatives have been driven by China's recent adoption of the International Organization for Standardization (ISO) 14000 international certification system,[82] coinciding with its entry into the World Trade Organization (WTO). By 2001, over 900 enterprises in China had been awarded the certificate.[83]

In addition, as a means of pushing environmental reform on the basis of consumer choice, the government is in the process of developing an environmentally safe labelling scheme. The Ministry of Agriculture introduced the green food label (that refers to food that has been grown under strictly controlled chemical use) in 1990. This was followed by the 'Three Green Project' initiated by SEPA in 1999 to 'cultivate a green market', to 'promote green consumption' and to 'provide a means for improving food quality'. China's labeling scheme is now more in line with international standards, although monitoring capacity is still weak, and national standards for organic food have only recently been introduced.

It is now highly likely that the Chinese government will continue to strengthen market incentives as a way of dealing with its environmental problems. In the absence of a fully functioning market economy, progress is likely to be slow and more successful in the developed coastal regions of China that can afford to pay. But even in China's richer regions, the question that needs to be addressed is whether it is possible to speed up the restructuring of industry while 'internalizing' environmental costs at the same time.

Clearly, government-driven market reforms have their limitations. Future progress will require the support and initiative of corporations. Chinese as well as foreign enterprises have usually been associated with contributing to higher levels of pollution rather than efforts to protect the environment. Yet, growing evidence suggests that corporate attitudes may be changing.[84] Some foreign enterprises have been actively involved in the environmental goods and services sector; others cooperate with Chinese manufacturers to produce environmentally friendly products or to promote energy efficiency.[85] An important area of future research will be to determine the extent to which corporations in China are taking the environmental initiative in response to consumer demands rather than simply reacting to government regulations.

Promoting public participation in environmental management is a more complex and politically sensitive task because it is highly contingent upon broader developments within Chinese civil society. The question of the extent to which China is experiencing the rise of a civil society (*gongmin shehui*) remains the subject of intense scholarly debate.[86] In the wake of China's economic reforms many scholars have sought to find evidence of an emerging Anglo-American style liberal civil society.[87] Others have proposed that associational life in China is 'state corporatist'.[88] In other words, social organizations are dominated by the state and are simply mechanisms for channelling state interests. But although 'state corporatism' explains an important dimension of civil society in China, the current reality appears to be more ambiguous. Moreover, the corporatist perspective overlooks the benefits that can be gained from engagement with the state. An *engaged civil society* has a greater potential to reshape the state and arguably in time provide an impetus for greater democratic freedoms. As noted by Tony Saich, Chinese NGOs stand to benefit from a close association with government in that they are better able to expand their influence and authority by 'working through government connections'.[89]

Clearly understanding the changing nature of state–society relations in China will require ongoing empirical investigations. In the meantime, it is perhaps fair to say that it is the medium through which citizens can have their say that is changing more dramatically than the political space in which they operate. Traditionally, public environmental concerns have been expressed on the basis of a letter (*xinfeng*) or visit to the local EPB.[90] Many farmers have also staged protests.[91] More recently, citizen hotlines (installed at the national, provincial and municipal levels) have become popular. The Internet also provides a new means of articulating protest as well as a source of ideas to facilitate government decision-making. In the planning stage of the tenth Five-Year Plan, the NDRC is reported to have received over 10,000 online suggestions from citizens (many of which were environment related). Apparently, more than 300 were adopted. In the words of one official from the NDRC:

> [The Internet] is a useful form in which to give fuller scope to democracy; people's active participation made the plan ever more practical and scientific, which is conducive to turning the implementation of the plan into a conscientious action of each citizen.[92]

The most important means through which Chinese citizens can voice their opinions is through the media. The Chinese media play a critical role in stimulating public concern as well as reporting egregious environmental damage. The whistle-blowing function of the media has become the *sine qua non* for ensuring that government efforts to protect the environment have a chance of success. Environmental media attention ranges from reports on the Baiji Dolphin in the Yangzte River and road construction in primary forests in Yunnan to Liu Xiaoqing (a famous Chinese movie star) collecting garbage. China Central Television (CCTV) has led a mass campaign on environmental awareness through the broadcast of its environmental programme '*Love our mountains and rivers, cherish the scenic spots and historical sites*'. *Renmin Ribao* has a column titled '*Touch water and turn it into gold*' that encourages ideas from ordinary citizens for conserving water resources and *Zhongguo huanjing bao* (China Environment News) provides monthly accounts on environmental issues and raises public awareness of the need for conservation. The media have also played an important role in monitoring environmental conditions. By the end of 1998, 59 cities had released urban air quality reports via the media on a weekly basis.[93] These reports have resonance: home buyers in Beijing are now reluctant to make investments in areas where the air quality is consistently low – reportedly leading to significant reductions in the value of housing situated in polluted areas.[94]

Clearly, the Chinese media are an important mechanism for channelling public opinion and raising environmental awareness but they too have their limits. The media cannot compensate for the development of autonomous social organizations; and they are no substitute for active public participation in environmental management. The problem is that Chinese regulations on the formation of social organizations (*shehui tuanti dengji guanli tiaoli*) are highly restrictive. First, before registering with the Ministry of Civil Affairs (MoCA), they must be sponsored by a government-owned unit; second, only one social organization is permitted for any single sphere of activity; third, they can only operate at the administrative level at which they are registered; and, fourth, they must have at least 50 individual members with initial assets of US$12,000.[95]

Despite these restrictions the number of registered environmental NGOs in China continues to rise. According to the Not-for-Profit Organization (NPO) network in Beijing, there were over 50 environmental NGOs operating in China in 2003.[96] Many remain under government control and are usually referred to as government organized NGOs (GONGOs) (*ban guanfang zuzhi*). These include the China Environmental Science Foundation, the China Wildlife Conservation Association and the China Environment Protection Fund.[97]

At the other extreme, a growing number of Chinese environmental advocacy organizations (*minjian zuzhi*) now operate with a measure of independence. Most attention to date has focused on a small number of high profile NGOs in Beijing such as Friends of Nature (*ziran zhi you*) established by Liang Congjie in 1994 and Global Village (*Beijing diqiucun jingwenhua zhongxin*) set up by Liao Xiaoyi in 1996. But many more independent NGOs have been set up in the regions, especially in Yunnan, Sichuan and Qinghai, benefiting from a concentration of donor funding and the support of transnational networks.[98] Environmental volunteer associations that are mainly attached to universities such as *Lü Jiayuan* (Green Homeland) at Jiaotong University in Beijing, or *Huanbao Ziyuan Xiehui* (Environmental Protection Volunteer Association) at Chengdu United University are also relatively independent. In addition, many Chinese citizens promote individual causes such as the Tibetan Antelope, or the Saunders' Gulls on the northeast coast of China.[99]

China's more independent NGOs are at the vanguard of green advocacy through their efforts to promote environmental education, habitat conservation and recycling.[100] Public participation is central to their work. The fact that participation has now been officially endorsed as a guiding principle in national poverty alleviation efforts means that the concept is becoming more familiar and easier to promote.[101] For example, an environmental NGO in Beijing, Community Action, dedicates its whole mission to advancing public participation in community development.

The fundamental importance of environmental NGOs, however, lies in their potential to play a critical role in nurturing the public sense of responsibility; they have an important mediating role to play in overcoming public passivity and breaking the dependence of the Chinese people upon the state to solve environmental problems. Breaking this dependency, in a psychological if not in a material sense, presents a major challenge. Under the central planning system, the Chinese people were long accustomed to the government provision of social welfare services ranging from housing and jobs to education and the environment. During the reform period the state provision of basic welfare services or the so-called 'iron rice bowl' is gradually being faded out, yet, in the eyes of many, protecting the environment remains a priori a government responsibility. This is a point to which I will return in Chapter 6.

To summarize, we have seen that state environmental management in China is more than simply a functional appendage of economic policy; it is comprehensive and even innovative in its embrace of economic incentives. But further progress is limited so long as the government remains wedded to the promise of an 'engineering fix', and is more concerned with imposing regulatory control than building capacities on the basis of enhanced economic efficiency and participation. As the environmental agenda becomes more complex, the issue of sufficient capacity to implement reform will become more critical. To this end, it is imperative that investment and resources are channelled into developing the legal system, establishing economic incentive schemes, and increasing public participation.

A more immediate problem lies in the responsiveness of local governments. Above all, the ability of the central government to deliver environmental improvement in China depends upon the political will and capacities of local governments. The 1979 Environmental Law established a legal basis for local autonomy over environmental management. In other words, the main responsibility for implementing environmental polices and regulations and thereby ensuring environmental quality lies with the local people's government. Over two-thirds of planned investment in environmental protection in China is now undertaken by local governments and enterprises.

But instead of serving as catalysts for environmental action, as in the case of Japan and the United States, local governments in China are seen as posing a serious obstruction. In effect, it would seem that a disjuncture exists between a central commitment towards environmental protection in Beijing and dissent in the provinces that presents the single biggest constraint upon solving environmental problems in China. The discussion that follows seeks to define the main conflicts of interest involved in local environmental management and to address the question of whether the problems are only related to weak compliance or also to weak capacities.

Managing the environment from below

The traditional tug-of-war between the state and its localities is a characteristic feature of Chinese history. Even in the post-1949 era, Mao Zedong stressed the need for 'local self-reliance' and therefore the scope existed for communes to pursue local interests, albeit heavily circumscribed under party controls.[102] The growing fragmentation between the 'Centre' and its localities in the post-1979 reform era has been well documented.[103] Jean Oi has suggested that local governments now behave more like 'local states' that 'coordinate economic enterprises in [their] territory as if [they] were a diversified business corporation'.[104] Municipal governments invest in local industry, have control over collective enterprises, and compete with other localities for foreign investment. Under these conditions, it is hardly surprising that market reforms have led to the profligate pursuit of economic self-interest at the cost of the community's environmental health.

Decentralization as a double-edged sword

At the local level in China the enforcement of environmental regulations is haphazard. EPBs are weak in relation to local production-oriented agencies and suffer from an inherent conflict of interest: they are responsible to the central environment agency SEPA in Beijing and also form part of the local government machinery. On balance, the power of local governments prevails because they provide environment agencies with their annual budgetary funds, and they also have decision-making authority over the allocation of personnel and resources. A vicious circle exists whereby EPBs are dependent for regulatory enforcement upon local governments which, in turn, depend upon local enterprises as an important source of tax revenue and employment.[105] To make matters worse, the already heavily indebted banking sector is unwilling to provide loans for environmental protection purposes because of the low return on investment associated with environmental projects.

The situation is made more complex because of agency capture or the 'self-regulation' problem.[106] Local leaders lack incentives to enforce national regulations because they often own or have strong vested interests in the enterprises that they are supposed to regulate. A good illustrative example of such government collusion is the poor performance of EIAs. Until 1989 a mere 0.1 per cent of regional construction projects were prevented on environmental grounds, with only 0.5–1 per cent cancelled in the 1990s.[107] Further conflicts of interest are generated by the expanding entrepreneurialism of EPBs seeking to provide consultancy services to the industries that they are responsible for regulating.[108]

Generally speaking, as observed by Gordon White, local cadres are strongly motivated to pursue 'self-preservation' through informal 'clientalism'.[109] The rise of the technocratic cadre in China is helping to move the system more in the direction of meritocracy and professionalism, but social practices are still heavily embedded in personalistic modes of practice.[110] A study by Hon Chan *et al.* on environmental control in Guangzhou provides an illustrative example of the difficulties imposed by China's informal and personalistic authority structure: EPB officials were unable to monitor emissions from one particular factory because the factory director had a higher administrative rank than the director of the EPB.[111]

The Central government has attempted to address the lack of incentives for environmental protection at the local level by adopting the responsibility system (*huanjing baohu mubiao zeren zhi*) whereby poor performance ratings in meeting specified environmental targets can adversely affect political careers.[112] In the words of the former President Jiang Zemin, 'when a cadre's political record is judged, family planning and environmental protection should be important factors'.[113] According to various media accounts, the responsibility system has had some effect in galvanizing local mayors into environmental action.[114] In 1999, SEPA initiated a new policy for EPB directors to be appointed by the central government.[115] But it remains uncertain as to whether central authority over EPB leadership will further create or reduce tensions.

All the factors discussed earlier amount to a double-edged sword: local management of environmental problems is essential because of the need to deal with local variance but decentralization is leading to an adverse effect. As the Chinese saying goes '*yuan shui jiu bu liao jin huo*' (distant water is useless in fighting the fire at hand); in this case it seems that water nearby is equally useless.

The importance of local capacity

The previous discussion conveys a strong impression that non-compliance at the local level is the sole cause of the failure to implement environmental protection in China – a view that is commonly reinforced by the central government. For example, the Director of SEPA, Xie Zhenhua, claimed in 1994 that 'what is most serious is not pollution but the wrong attitudes of local authorities'.[116] Yet, such a perspective suffers from too abstract a level of generalization. On closer examination, it appears that local environmental management is not a uniform failure. Instead, performance is mixed and uneven across different regions. Indeed, in the more affluent cities on the east coast demands for a cleaner environment are relatively high.

The discharge levy system now accounts for approximately 25 per cent of the salaries of all national environmental staff members.[117] Hence, a strong incentive exists for environmental agencies to enforce environmental standards. A Study by Wing-Hung Carlos Lo and Shui-Yan Tang suggest that EPBs have had some success in controlling industrial pollution, although TVEs still remain largely outside of administrative control.[118] Moreover, in the face of resistance from polluting enterprises some environmental agencies have taken the initiative and adopted new innovative practices.[119]

The degree to which regions in China differ in their environmental performance is more difficult to evaluate. Empirical studies are still limited both in volume and geographic scope. The majority of studies focus upon the richer eastern provinces and municipalities – namely Guangzhou, Shanghai and Jiangsu.[120] However, it would seem that the uneven spatial distribution of China's economic growth is leading to a situation whereby the ability to act environmentally will be higher in the more developed regions of China where local government financial and institutional capacities are stronger. It is, perhaps, for this reason that in 1995 Shanghai had 17 water treatment plants whereas Guizhou had only one.[121]

Beyond the issue of compliance, as hinted at by Zhang Weijong *et al.* and Lo and Tang, weak capacity in environmental management is also a key variable in determining the success of local implementation.[122] Local officials are not necessarily exclusively motivated by self-interest – if they are, the consequences for the environment in China are dire indeed, and international environmental aid would represent little more than a leap of faith. Other motivating factors such as professional pride and satisfying community concerns are also likely to be important. From a broader perspective, local environmental action is likely to vary according to the funding, skills and institutional capacities of the EPBs, the political commitment of local leaders and the level of public awareness.[123]

Implications for international environmental aid

This analysis of the nature and scope of environmental management in China suggests two clear implications for international environmental aid: first, that the potential does exist for encouraging the Chinese government to broaden its approach to dealing with environmental problems; and second, that building capacities at the local level is likely to be the most effective means of improving local environmental management by providing a counterbalance to the problem of non-compliance.

The central government's recent decision to test the use of market incentives for environmental purposes is a positive signal. Nevertheless, the promotion of a market-oriented approach to environmental management is likely to present a number of difficulties in the current climate of industrial restructuring. The scope for promoting a participatory approach is more limited. Donors not only face political constraints but also an attitudinal barrier on the part of the Chinese people long accustomed to the government taking sole responsibility for the environment.

Terminology and differing conceptions also present a problem. Whereas the notion of incentives is easily understood in China (at least in theory if not in practice), the notion of participation is prone to misunderstanding. The contemporary usage of *qunzhong canyu weiyuanhui* to denote participation has strong communist connotations and literally means mass-based committees. In more recent official discourse, participation has been translated as *zhengfu zhudao, gongtong canyu* which literally means 'joint participation under the guidance of the government'.[124] The Western concept of stakeholder participation is especially difficult to translate into Chinese. It can best be translated as *huanbao canyuzhi*, meaning broad-based environmental associations. Confusion over terminology can lead to a situation in which the notion of participation as expressed by donors is at odds with the understanding of the recipient.

Donor assistance in China is needed to reinforce the relevance of both economic incentives and participatory practices. The most important role for the international donor community, however, lies in using these approaches to build environmental capacity at the local level. This is no easy task for two important reasons. First, although the Chinese government has made considerable efforts to develop both a long-term environmental investment plan and a comprehensive framework for environmental regulation and legislation, virtually no attention has been given to developing broad-based environmental capacity. Donors, therefore, need to convince the central government not only of the importance of using non-state approaches for managing environmental problems, but also of the need for capacity building.

Second, many conflicts of interest are involved in environmental management at the local level. To win support for building capacity, donors need to demonstrate the environmental *and* the economic benefits of environmental protection. To this end, environmental projects require in-built training provisions, incentives and participatory practices. The advantage that donors have is that the prospect of funding per se can enhance political commitment. The implementation of internationally funded environmental projects is fundamentally different from the implementation of environmental regulations and standards. Local governments stand to benefit more from the former, both in terms of funding and status. According to the latest figures available, by 1999 China had received approximately US$3.6 billion in loans and US$420 million in grants from foreign countries and international organizations for environmental projects. In addition, the Global Environment Facility had provided US$210 million (19 projects in total) and the Multilateral Fund of the Montreal Protocol an additional US$275 million (266 projects in total).[125] Most of this funding went directly to the local governments that were also primarily responsible for the repayment of the loans.

Conclusion

The task of environmental management in China is formidable. China's environmental problems are immense and difficult to solve in the current climate of socio-economic transition. Despite the difficulties involved, this chapter has argued that the Chinese government now has a strong commitment to environmental protection. Although it has a preference for engineering solutions, more recently it has been willing to address the underlying causes of environmental degradation by experimenting with economic incentives.

The problem is that the initial impetus for a more incentive driven approach to environmental management has not been matched with a sufficient interest in the large-scale promotion of participatory practices. Consequently, a shared sense of environmental responsibility between the government and the Chinese people is largely absent and limited to a small but growing number of pioneering NGOs. In an era when China is striving to replace centralized planning practices with individual incentives, the notion of collective or shared responsibility seems old-fashioned. Yet, if China is to solve its environmental problems before it becomes more democratic, which could take decades, it needs to actively promote both economic incentives and public participation.

One of the biggest problems that China faces in the immediate term is the uneven nature of government compliance at the local level. The implementation of environmental policies and regulations has been fraught with conflicts of interest. On closer examination, however, it would appear that weak compliance is not the only impediment to improving local environmental management; weak capacities are also likely to be a major constraint. Under these circumstances, the international donor community has an important role to play in both encouraging a comprehensive approach to environmental management in China and in building local capacity. The fundamental challenge for donors is to work effectively with both central and local actors involved in the design and implementation of international environmental aid. In order to do so, environmental donors need to steer a narrow course between the Scylla of political motivation and the Charybdis of economic self-interest.

In the following three case studies we shall see that donors have had some success in convincing local governments of the need to take into account the longer term costs of unchecked pollution and unsustainable practices by implementing new approaches. Many local officials have been easier to convince than their central counterparts of the need to invest in capacity building. Yet, donors are by no means united in pushing for environmental reform in China, particularly at the level of actual project implementation. An attempt to understand the effectiveness of international assistance will also reveal some interesting paradoxes in donor environmental practice.

3 Engineering a solution

The Japanese approach

> It is the striving of every individual in a society based on a belief in the future that is the most powerful driving force behind nation-building. The foundation of Japan's development was its 'ownership' of that development – that is to say, we undertook our own development with our own efforts.[1]
>
> (Shigeo Uetake, Japanese Senior Vice-Minister for Foreign Affairs)

In some ways, this first case study on Japanese environmental assistance to China is the most straightforward. The Japanese belief that environmental problems can best be solved through improved technologies and engineering capability meshes comfortably with the orientation of the Chinese government. Japan's own experience in dealing with urban pollution problems during the 1960s is also relevant to the current conditions in China. Yet, to a large extent, Japanese environmental assistance to China is shaped by the nature of the bilateral relationship. As a result of historic political sensitivities, Japanese aid policy dialogue with China is still constrained, making it difficult for Japan to impose its own environmental agenda. Consequently, the Japanese approach to solving environmental problems in China is largely technocentric with a strong emphasis upon engineering solutions. Capacity building is targeted to the promotion of environmental research and training on the basis of grant assistance.[2]

The Japanese predisposition towards infrastructure building and technical fixes has been heavily criticized by Japanese and Western scholars.[3] In the words of Matsuura Shigenori, 'the Japanese government and industry are more interested in profits from the sale of environmentally safe products than in planetary salvation'.[4] Such accusations are often unwarranted. At least in the case of China, the commercial benefits from Japan's environmental assistance have been slow to materialize. A more serious problem with such critiques is that they do not take into account the needs and priorities of the recipient. For China, given its specific conditions, there is a clear advantage to be gained from receiving large-scale financial transfers for environmental infrastructure. Many cities in China do not have efficient sewerage or waste disposal systems and water supply and heating systems are often archaic. If left to the private sector, it is highly unlikely that these environmental infrastructural needs would be met because the return on investment in the short term is typically low.

When assessing the effectiveness of Japanese environmental assistance it is therefore important to take a more neutral stance. Even if the prospects for building environmental capacity may at first appear remote, it is still essential to identify where future improvements can be made. The critical questions are three-fold: to what extent is an engineering approach to environmental management compatible with environmental improvement? Is the Japanese approach, despite its engineering focus, still in some way effective in building local environmental capacity? And, if this is the case, under what conditions is it effective?

The findings in this case study suggest that Japanese environmental assistance is leading to a visible reduction in pollution loads in some Chinese cities. It is also well received by local officials in China with the advantageous effect, in poorer regions, of increasing local political commitment towards the environment. Moreover, under certain conditions, the Japanese approach has had a positive effect by bolstering local capacity for financial management. Contrary to the expectations of some scholars,[5] what is surprising is that the approach has not had any significant impact upon local technological capacity.

On a less positive note, serious questions remain over the long-term benefits of Japanese aid, especially in relation to industrial pollution control. Large-scale environmental funds require more careful monitoring and sophisticated environmental impact assessment methods. They also need to be carefully coordinated with complementary investment in broad based capacity building – technological, financial, institutional and social. It is in this sense, rather than in infrastructure building per se that Japanese environmental assistance falls down.

Addressing this problem would require more grant assistance and better coordination between the Japanese grants and loans agencies, the Japan International Cooperation Agency (JICA) and the Japan Bank for International Cooperation (JBIC) respectively. Notwithstanding the perennial problem of bureaucratic resistance, a major obstacle to realizing this goal is Japan's development philosophy. The experience of post-war reconstruction has been fundamental in shaping the nature of Japanese aid. It is precisely the firm belief that the solution to development lies in the self-help efforts of developing countries themselves that distinguishes Japan from other Western donors. It is also this belief that reinforces the Japanese preference for loans rather than grants. In the words of Kusano Atsushi:

> From the Western perspective aid fundamentally means extending grants in the spirit of charity...and does not mean loans for which repayment is sought. The principle behind Japan's aid, however, is to help developing countries help themselves. It is thought that the developing countries receiving yen loans would make such self-help efforts as they draw up repayment plans, in other words, they begin to build up their own nations.[6]

For Japanese aid agencies capacity building is equated with 'ownership'. In relation to the environment this means that 'developing countries assume the primary responsibility and roles to address environmental issues', and that 'donor countries assist such self-help efforts'.[7] But if Japan is to seriously improve the effectiveness of its environmental assistance then it needs to move imperatively beyond the issue of 'ownership' and towards broad-based capacity building.

This chapter will trace the implementation of Japanese environmental assistance from the boardroom in Tokyo to the municipality in China. In so doing, it will identify donor and recipient interests and priorities, and analyse the local responses to Japanese environmental loan projects in Shenyang, Benxi, Huhhot, Liuzhou, Dalian and Guiyang. The focus of concern is with environmental loans rather than grants because the former account for over two-thirds of total Japanese environmental assistance to China and are largely concentrated on pollution control. Proceeding from the project investigations, the chapter will conclude by assessing Japan's contribution to improving local environmental capacity.

Reorienting Japanese aid towards the environment

In 2001, after a period of 11 years, Japan lost its status as the world's largest aid donor to the United States. At the same time as Japan decreased its aid budget by 10.3 per cent to US$9.6 billion, the United States increased its aid budget by 9.3 per cent to US$10.8 billion.[8] The Japanese government was forced to cut its ODA in response to tighter fiscal constraints and increasing public scrutiny over its aid policies. Aid to China was singled out for special attention leading to a 25 per cent reduction in the planned budget for 2001. This represented the first reduction in aid to China since the first ODA package was negotiated in 1979. Chinese concerns were allegedly assuaged by the fact that this could be justified solely on the grounds of Japan's ailing economy.

But in reality, public scepticism of Japan's ODA to China runs deep. It is perpetuated by the sentiments of a younger cohort of politicians who, unlike their elders, do not feel that they have a moral debt to pay for Japan's past aggression.[9] Criticism is generally motivated by three key concerns. First, China is Japan's second largest recipient of aid, receiving approximately 200 billion yen annually, yet it is generally felt by many Japanese that China has now graduated from its former dependency upon foreign aid. Second, the Japanese public is deeply suspicious that its ODA has somehow been linked to increases in Chinese military spending.[10] Third, and perhaps most importantly, there exists a common feeling amongst the Japanese public that its ODA is not appreciated in China.

The public furore over aid to China, that reached its height in 2000, was double edged in that it coincided with broader concerns over Japan's aid administration. Negative public sentiment soared when a number of scandals in the foreign ministry unveiled the extent to which ODA funds were being misappropriated by senior Diet members and bureaucrats. The government responded by setting up a new 18-member ODA reform council headed by the foreign minister Kawaguchi Yoriko. However, the decades old debate over the need for a single development aid agency remains deadlocked. Even in the current climate of reform, the government has tended to sidestep the issue in the face of fierce bureaucratic resistance.

Japanese aid decision-making is complex.[11] More than 15 bureaucracies are involved in the process with three main ministries vying for dominant influence – the Ministry of Economic Trade and Industry (METI),[12] the Ministry of Foreign Affairs (MoFA), and the Ministry of Finance (MOF). MoFA has jurisdiction over Japan's grants agency, JICA (*kokusai kyôryoku jigoyô dan*),[13] and both METI and MOF preside over Japan's loans agency, JBIC (*nihon kokusai kyôryoku ginkô*). JBIC was formed in October 1999 when the Overseas Economic Cooperation Fund (OECF) merged with the Japanese Export and Import Bank (JEXIM).[14] The present case study was conducted both before and after the merger. To avoid confusion I shall refer throughout this chapter to JBIC.

The complexity of the Japanese aid administration makes it difficult to define Japan's donor interests. Scholars have offered different perspectives ranging from foreign strategic imperatives, to bureaucratic politics, or economic cooperation.[15] While all these factors have been important in shaping Japanese ODA policy, it would seem that economic interests have been paramount. Historically, METI, in representing government and corporate leaders, has taken the leading role in shaping Japan's development assistance, otherwise known as economic cooperation (*keizai kyôryoku*).[16] This was further reinforced in the 1990s by growing domestic fiscal constraints together with the onset of the Asian financial crisis in 1997. Tokyo's economic interests remain prominent with aid now being re-linked to key markets for Japanese exports.[17] But ultimately, Japan's ODA needs to satisfy foreign policy as well as economic interests. As a non-military power, Japan is reliant upon its ODA for political and diplomatic purposes. This is because ODA provides an important means of contributing to international peace and security efforts as well as strengthening Japan's bilateral relationships with developing countries.

Over the past decade, reorienting aid towards the environment has provided the Japanese government with a neutral means of reconciling its competing economic and political interests. As an energy resource-poor nation that relies upon imports for more than 80 per cent of its primary energy requirements, Japan is keen to promote energy conservation amongst its neighbouring countries in order to reduce the risk of regional competition for energy supplies.[18] Japan is also home to some of the world's most competitive environmental industries.[19] At the same time, as many commentators have noted, environmental aid is central to Japan's ambitions to play a role in the international community that is commensurate with the size of its GDP.[20] Throughout the 1990s, Japan actively participated in international environmental conferences and made significant contributions to global environmental funds. Despite the recent reduction in Japan's overall ODA budget, environmental aid has remained a core priority. At the World Summit on Sustainable Development in Johannesburg in 2002, Prime Minister Junichiro Koizumi pledged that 'Japan will continue to extend its environmental cooperation, mainly through its ODA, in order to support sustainable development in the world'.[21]

The scope of Japanese environmental aid

Japan is now the world's largest bilateral donor of environmental aid. Since the late 1980s, Japan has repeatedly targeted the environment as a major component

Table 3.1 Japanese aid in the environmental sector 1990–1999 (US$ million)

Fiscal year	Grant aid	Technical assistance	Loans	Multilateral assistance	Total
1990	161.4	93.6	879.0	34.8	1,168.8 (12.4)
1991	181.3	106.0	500.5	58.2	846.0 (7.2)
1992	249.1	139.6	1,774.3	84.8	2,247.8 (16.9)
1993	349.8	187.5	1,416.3	150.3	2,103.9 (12.8)
1994	405.4	229.2	1,033.0	247.8	1,915.4 (14.2)
1995	455.2	237.0	1,815.9	425.5	2,933.6 (19.9)
1996	331.4	232.8	3,551.4	141.3	4,256.9 (27.0)
1997	298.0	248.5	1,341.6	130.6	2,018.7 (14.5)
1998	289.9	304.2	3,280.8	263.1	4,138.0 (25.7)
1999	250.5	240.9	3,961.8	116.0	4,569.2 (33.5)
Total	2,972.0	2,019.3	19,554.6	1,652.4	26,198.3

Source: Compiled from Japan, Ministry of Foreign Affairs, *Japan's Official Development Assistance Annual Report*, Tokyo: Association for the Promotion of International Cooperation, various years 1996–2000.

Notes
a Amounts are calculated on a commitment basis for grant and loan assistance, on a disbursement basis for technical cooperation, and on a budget basis for contributions to multilateral institutions.
b The figures in parentheses in the total column represent the share of total ODA commitments for that year.
c Dollar amounts are based on conversion exchange rates on a year on year basis. The reduction in environmental aid from 1996 to 1997 appears greater when converted into US dollars due to exchange rate fluctuations. The dollar rate declined from ¥108.82 to the dollar in 1996 to ¥121 to the dollar in 1997.

of its ODA. At the G7 Arche Summit in 1989, Japan made a commitment to spend 300 billion yen (US$2.17 billion) on the environment globally. Three years later at the United Nations Conference on Environment and Development, Japan made a further pledge to increase its environmental aid to US$7 billion by 1995. Actual spending surpassed US$9 billion. Despite the fact that Japan's total ODA budget fell from US$14.7 billion in 1995 to US$8.6 billion in 1998,[22] environmental aid continued to grow. From the early 1990s onwards, Japan provided approximately US$2 billion per annum in environmental aid, reaching a cumulative total of over US$26 billion by 1999 (see Table 3.1).

Japanese environmental aid is defined broadly to cover a wide range of projects relating to reforestation, disaster prevention, pollution control, natural environmental conservation and capacity building in environmental policy and institutions.[23] Between 1990 and 1996 environmental aid was predominantly in the form of grants[24] and multilateral assistance. Japanese environmental grants began in the mid-1980s and largely targeted water supply and forestry development. In the 1990s the focus then shifted to environmental research and information centres.

It is significant that during this period the sectoral distribution of environmental aid strongly reflected the interests and needs of the recipient countries. Statistics for 1997 show that 60 per cent of total bilateral environmental aid was directed towards living environment concerns (water supply and sewerage) and disaster prevention, with the remainder allocated for pollution control and conservation. This changed dramatically following the introduction of environmental loans in fiscal 1996. These were largely focused upon industrial and municipal pollution control and constituted 25 per cent of total aid loans.[25] By 1998, over 60 per cent of bilateral environmental aid was focused upon pollution control. China was one of the first recipient countries to benefit.

It is important to acknowledge, however, that Japan's active leadership role in environmental assistance remains strongly dependent upon the support of recipient countries. Many developing countries are cautious about accepting vast amounts of environmental aid. It is often seen as a form of indirect protectionism to further benefit the economic competitiveness of the richer countries that are better placed to deal with their environmental problems. To allay these concerns, Japan has used its own experience in pollution control during the 1960s as a way of providing the necessary green (or rather brown) credentials to counterbalance recipient country suspicions over Japanese motives.

The Japanese experience in pollution control

In the 1960s, Japan learned valuable lessons in regard to the economic and social costs of ignoring pollution control. The now famous case of Minamata disease – when a polyvinyl chloride (plastics) factory dumped untreated waste products into the Minamata bay, leading to widespread mercury poisoning – has been estimated to have cost industry approximately ¥200 billion in compensation payments to the victims.[26] If the company in question had taken preventative measures against mercury effluents, it would have cost an estimated ¥1.5 million.[27] The case is still not closed and the social costs to human lives are immeasurable.

Japan continues to face environmental challenges *vis-à-vis* hazardous waste disposal, recycling and waste minimization, ecological destruction along its coastline, and transportation. The overseas investment activities of Japanese firms have also led to considerable environmental deterioration.[28] Like most countries, therefore, Japan is in no position to provide a perfect model of sustainable development. However, in relation to urban environmental management, Japan's early experience in pollution control provides important lessons for other densely populated and rapidly industrializing developing countries. In particular, the Japanese government has not been slow to point out its own historic environmental failures and their significance to China, as articulated by Yoda Susumu (President of the Central Research Institute of Electric Power Industry in Japan), at a Conference in Beijing in 1996:

> From Japan's point of view, the regrettable thing is that China will duplicate Japan's mistakes, more or less, and pollute its air and water as a result of achieving economic growth so rapidly. In our history we Japanese have learned much from the mistakes that have caused us pain.[29]

The Japanese experience of controlling pollution during its period of rapid economic growth in the 1960s and 1970s has four main characteristics: first, Japanese industry was able to rationalize production costs and invest in pollution control at a time of increasing world oil prices; second, central government provided the necessary financial support by providing tax incentives, low interest finance to local government and low interest loans to industry financed through a pollution control corporation; third, local governments were at the vanguard of environmental protection and had a significant degree of autonomy in setting environmental standards and strengthening national regulations; and fourth, citizens' movements played a crucial role in placing environmental issues on the political agenda and encouraging industry compliance.

Above all, the Japanese experience revealed the need for cooperation between government (central and local), industry and citizens in order to resolve environmental problems. Public participation was indispensable. Citizens' movements, triggered by individual concerns and public outrage over serious outbreaks of pollution-related diseases, were important in bringing about the introduction of pollution control agreements.[30] These agreements between local government and industry led to the development of voluntary emissions controls that proved to be highly effective. According to one of Japan's leading experts, Dr Michio Hashimoto, the strength of the citizens' movements lay in the fact that they were supported by professionals with the necessary scientific evidence to gain leverage over local governments and industry.[31]

How relevant is the Japanese experience in urban environmental management for developing countries? The Japan Environmental Corporation (JEC) (formerly known as the Environment Pollution Control Service) provides a useful institutional model for stimulating enterprise investment in pollution control, and many of the end-of-pipe technologies and cleaner production processes adopted by industry can be replicated in other countries. But Japan's success in pollution control was also reliant upon a market mechanism to induce competition. The oil shocks of the 1970s motivated electricity price reform in Japan in order to reflect the economic cost of supply. According to the Environment Agency in Japan, the introduction of cleaner production processes stimulated by energy price reforms was more significant in reducing SO_2 emissions than end-of-pipe desulphurization measures.[32]

The ability to enforce stringent environmental standards at the local level was also dependent upon a political balance of power between citizens, local governments and industry. Effective environmental action was primarily rooted in the fact that local leaders had been elected and therefore had an obligation to take into account public concerns over pollution. In addition, the success of the citizens' movements depended upon an established judiciary system. The victims of pollution diseases sued the industries responsible for damages suffered and the courts eventually ruled in favour of the plaintiffs.[33] Although no legislative framework existed (the Basic Law for Environmental Pollution Control was not enacted until 1967), the conditions were right for successful implementation: Japan had an established market economy with well-developed democratic institutions.

The Japanese experience provides a clear example of the need for an integrated approach to environmental management that involves regulations, incentives, technological innovation and public participation. But the experience has had a minimal impact on Japan's actual approach to solving environmental problems in China. Voluntary agreements or utility price reform are not seen as feasible in a transitional authoritarian country such as China. Hishida Katsuo, chief advisor to JBIC on environmental assistance to China, argues that local conditions in China largely dictate the Japanese approach to environmental management:

> Voluntary style agreements will not work in China and neither will market incentives. The idea of corporations producing a social good is not understood in China and the idea of social disclosure is also difficult; because Chinese industries do not feel the same sense of shame in regard to pollution as Japanese industries did in the 1960s. Market incentives won't work because companies cannot afford high taxes. Also cleaner production is very difficult to promote hence the emphasis upon end-of-pipe solutions.[34]

Although the transferability of the Japanese experience is deemed to be limited, the lessons still send a powerful message of 'prevention first' to developing countries. Above all, Japan has used its experience as an important diplomatic device for convincing recipient governments of the need to prioritize environmental concerns. This has been particularly important in the case of China, where Japanese aid agencies have tended to use the experience as a way of enhancing environmental dialogue with their counterpart agencies.

The greening of Japanese aid to China[35]

Japan has only recently placed a strong priority upon environmental issues in formulating its aid policy towards China. Historically, Japanese aid to China has centred upon yen loans for economic infrastructure, especially in the transportation sector.[36] Environmental projects, broadly defined, began in the late 1980s and focused largely upon water supply issues. The turning point came in 1991, when the first country study report on Japanese aid to China, under the chairmanship of Dr Saburo Okita (former foreign minister for the Ohira cabinet), recommended that the environment should become a priority sector.[37] The following year, Japan's environmental initiative in China was launched with an agreement to establish a Japan–China Friendship Environmental Protection Centre in Beijing.[38]

In 1994, environment-related projects comprised 15 out of the 40 projects to be funded as part of the first phase (fiscal year 1996–1998) of the fourth yen loan package.[39] The new environmental focus provided a means of weaning the Chinese off traditional infrastructure projects in the belief that China was now in the position to partly finance its own economic infrastructural needs.[40] But initially, despite its alleged concern for the environment, the Chinese government was not forthcoming in requesting environmental assistance from Japan. Consequently, Japan initiated a series of high-level policy dialogues as a means of placing its own preferences on the aid agenda. In 1995, a high-level environmental cooperation mission was dispatched to China under the leadership of the former Japanese Prime Minister Murayama Tomoichi. The following year the Japan–China Comprehensive Forum on Environmental Cooperation was held in Beijing, with participants from Japanese and Chinese local governments and the Japanese private sector.

Although ODA was central to environmental cooperation efforts, other conduits were opened via the private sector, in relation to METI's Green Aid of 1992, and the Japan Fund for the Global Environment.[41] Special emphasis was also placed upon local government cooperation, based upon established sister-city relationships. For example, Kitakyûshû has strong ties with Dalian in environmental protection while Hiroshima is linked to Chongqing through the establishment of an acid rain research centre. In 1996, representatives of Japanese and Chinese local governments met in Kitakyûshû City for a Japan–China Environmental Cities Meeting in order to exchange experiences.

The shift towards environmental cooperation during the 1990s culminated in the Sino-Japanese twenty-first century environmental cooperation agreement signed in November 1997. The agreement was preceded by Japan's then Prime Minister Hashimoto's visit to Beijing in September 1997, in commemoration of 25 years of resumed bilateral ties. The visit produced a number of concrete measures and financial initiatives to encourage further bilateral environmental cooperation. Hashimoto proposed a deepening involvement in environmental projects through the establishment of an environmental information network, together with an ambitious plan to facilitate the development of three environmental model cities. A joint Japan–China expert committee was set up to formulate detailed plans.[42]

The Japanese Prime Minister also announced a further increase in loans to improve the global environment (forestry, energy conservation and new energy sources) and to promote pollution abatement. Environmental loans were offered with softened conditions – annual interest rates were to fall from 2.5 to 0.7 per cent with a repayment period of 40 years and an extended grace period of 30–40 years.[43] Conditions were, therefore, comparable to World Bank concessional lending (the International Development Association provides grant assistance at no interest rate but with a 0.75 per cent service charge).[44] In fiscal 1998, an additional 79 loan projects were approved and almost half (32 projects) were environment related at a total cost of ¥322 billion (US$2.9 billion).[45] Environmental loans continue to constitute a major component of Japan's development assistance to China. In fiscal 2001, 7 out of 15 loan projects were environment related amounting to ¥88 billion

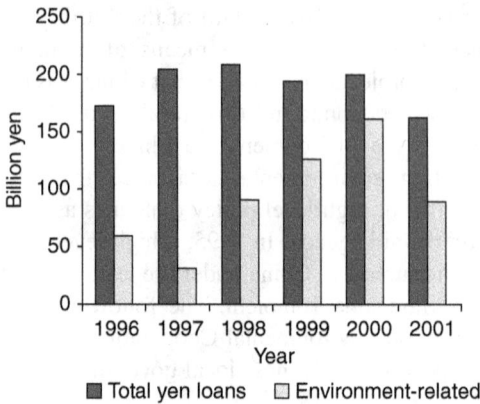

Figure 3.1 Japanese environmental loans to China 1996–2001 (in billion yen).
Source: JBIC, *JBIC Annual Report*, Tokyo: JBIC, various years 1999–2001.

(US$7.4 billion)[46] (see Figure 3.1). More recently, in the ¥96.6 billion loan package for fiscal 2003 environmental conservation projects accounted for 53 per cent of the total amount.[47]

Japan's present strategy is to focus its environmental assistance upon air pollution control in the three Chinese 'model cities' of Dalian, Chongqing and Guiyang.[48] In 1998, the second country study report on China, under the chairmanship of Professor Toshio Watanabe of the Tokyo Institute of Technology, stipulated that Japan should *save the best steel for the blade of the sword*. In other words, given the scale of China's environmental problems, Japanese environmental assistance should concentrate on severe problems that China is unable to alleviate on its own.[49] The new economic cooperation programme for China, launched in October 2001, continues to stress the importance of pollution control, especially in the lesser developed central and western regions of China. It also introduces new environmental priorities in the areas of information systems, renewable energy and ecological conservation.[50]

Japanese motives

The motives behind Japan's environmental aid to China involve a confluence of factors relating to economic pragmatism, bilateral diplomacy and regional/global environmental interests. Naturally China, with its burgeoning energy needs and vast potential for investment in environmental products and services is an obvious target for Japanese environmental assistance. But initially, as in the case of all JBIC loans to China, environmental loans were untied: the allocation of procurement contracts was not tied to Japanese firms but instead determined by a process of open international bidding.[51] Commercial interests seem to have been something of an afterthought. JBIC has only recently shifted towards a policy of partially tied[52] environmental loans in response to Japanese corporate frustrations over losing procurement contracts funded from Tokyo.[53] Moreover, according to officials at the Japanese MoFA, the private sector is becoming more interested in environmental investment in China.[54]

In providing environmental loans to China, political interests have tended to overshadow commercial interests. During the 1990s, environmental cooperation provided a neutral means of strengthening the bilateral relationship at a time of strained diplomatic tensions – triggered by Japanese objections over China's nuclear tests in 1995 and 1996, disputes over the Diaoyu islands in the East China Sea, and the 1997 review of the 1978 US–Japan Defence Guidelines. In the words of Mori Katsuhiko, professor in International Relations at the International University in Japan, 'the environment is one area where the Japanese government stands to gain a better reputation'.[55] Whether Japan can realize this objective remains to be seen given that Sino-Japanese diplomacy continues to be dogged by the issue of unresolved war enmities.

As Japan struggles to come to terms with its future, it equally struggles to come to terms with its past. When Prime Minister Hashimoto sought to promote environmental cooperation during his visit to China in September 1997, he was poignantly reminded that Japan had dumped chemical weapons in China during the Second World War. The irony could not be greater.[56] In 1999, a Japanese embassy official in Beijing defined Sino-Japanese environmental cooperation 'as long term, flexible and full of misunderstandings'.[57] Yet, Japan continues in its tireless efforts to break with the shackles of the past and engage China in a more vital and regionally oriented relationship, as illustrated in a speech by Hashimoto to the Japanese Diet in 1997:

> Cooperation between Japan and China must go beyond bilateral cooperation and be oriented towards international contributions to the entire Asia region and the world as a whole. For example, the environmental issue, as well as population and energy, are issues which impact not only on China but also on Asia and the entire world, making it necessary that Japan and China share one another's wisdom.[58]

The issue of acid rain provides the Japanese government with a practical means of shifting concerns away from beleaguered bilateral relations and towards regional cooperation. It is also congruent with Japan's own experience of dealing with high levels of SO_2 emissions in the 1960s – the first desulphurization facility in the world was constructed at a Japanese oil refinery, Idemitsu Kôsan, in 1967. Hence, Japan is in a strong position to offer financial and technical assistance. It would be misleading to suggest, however, that the acid rain problem is only of interest to Japan. More accurately, it is an issue of mutual interest – as indicated in Chapter 2, SO_2 levels in China are high and acid rain has caused far more visible damage in China than it has, as yet, in Japan.

Statistics on the origins of sulphuric ions detected in Japan continue to be shrouded in scientific uncertainty; both the *Kôsa* (a yellow sand transported to Japan from the deserts of the Asian continent) and Japanese forest soils have acid neutralizing capability.[59] Unofficially, according to a report carried out by the

Japanese Central Institute of Electric Power, it is estimated that 50 per cent of Japan's sulphuric ions are produced in China with 15 per cent in Korea, and the remainder in Japan.[60] On the other hand, some Chinese scientists have argued that acid rain in Tumen and Dandong (in northern China) comes mainly from Japan and Korea.[61] To help resolve the scientific uncertainties, a regional acid rain monitoring centre opened in 2000 involving ten countries in North and Southeast Asia – including China.

Japan also stands to gain from strengthening environmental cooperation with China at the global level. As a signatory to the Kyoto Protocol, Japan is required to reduce its emissions of CO_2 and other greenhouse gases by 6 per cent from 1990 levels by 2012. Under the Clean Development Mechanism (CDM) industrialized countries can earn credits towards meeting their emissions targets by providing financial assistance to reduce emissions in developing countries. At the seventh conference of the parties to the 1992 United Nations Framework Convention on Climate Change (or COP 7) held in Marrakech in November 2001, Japan proposed that the signatories to the Kyoto Protocol should be allowed to use ODA for CDM projects. Given Japan's large-scale lending in the energy sector, this proposal has not been well received by other signatories to the Protocol out of fear that it will simply lead to a re-labelling of current spending. Meanwhile, Japan has wasted little time in proposing a number of CDM projects to the Chinese government.[62]

The earlier discussion suggests that in the future strengthening environmental aid offers the most plausible means through which Japan will be able to maintain strong diplomatic relations with China as well as satisfy public concerns at home. Much will depend, however, upon effective implementation. Overcoming the difficulties involved in negotiating the environmental aid agenda is in itself a major challenge. How much influence does Japan have over the aid-agenda-setting process in China? And to what extent do Japanese project priorities meet Chinese needs and expectations? To address these questions, let us now turn to the process of implementation, starting with a discussion of environmental aid negotiations and moving onto an analysis of actual project performance at the local level.

Sino-Japanese environmental aid negotiations

The actual selection of Japanese loan projects is strongly biased towards Chinese concerns. This is because Japanese ODA is recipient-initiated (*yôsei shugi*) – recipient countries identify projects and formulate proposals to be approved by the donor. Japan does not have the political leverage to apply 'green conditionality' and therefore, relies upon persuasion through policy dialogue. As Ryukichi Imai notes:

> [Even] when Japanese ODA is used for the construction of thermal power stations, Japan cannot attach the condition that sulphur removal equipment be installed. ODA programmes are provided on the basis of a requirement of the Chinese government, and Japan cannot force China to adopt a clean air policy.[63]

Negotiations over the introduction of environmental loans were initially difficult and time-consuming. In the case of some loans, negotiations lasted for five years before an agreement was eventually signed.[64] In time, with a firmer commitment from the Chinese side, negotiations have become less contentious but tensions still remain.

Project priorities

From the outset, negotiations over environmental loans involved conflicts of interest in three main areas: the environmental focus of projects, the low rate of financial return on investment and environmental technology transfer. Characteristic of the traditional South–North divide over local environmental needs versus global priorities, the Chinese were initially reluctant to shift the traditional environmental focus away from water supply and sewage and towards air pollution. They considered environmental losses and health risks related to water conservation and pollution to be more immediate and serious than the effects of air pollution. Japan, on the other hand, had a keen interest in promoting air pollution projects in China because of its historic experience and expertise, together with concerns over acid rain.[65] Following protracted negotiations, a compromise was reached in which a number of so-called environmental improvement projects were introduced onto the agenda for the first time, including both air and water priorities. Projects were scheduled to commence in 1996, with each project broken down into a number of smaller sub-components (104 in total).

Contrary to the assumptions held by some commentators,[66] Japanese environmental loans were not directed towards reducing CO_2 emissions. Air pollution components involved the expansion of gas supply and central heating systems, dust control and the reduction of SO_2 emissions. By 1998, air pollution abatement had become the central focus of Japanese environmental loans. The rationale, agreed upon by a joint expert committee, was that water pollution has a direct impact upon the quality of life and is, therefore, more likely to be tackled by local governments whereas air pollution – particularly invisible emissions of NO_x and SO_2 – is more likely to be ignored.[67] More recently, environmental loans have concentrated on reducing air pollution by investing in natural gas supply and forestry conservation.[68]

Although a consensus has now been reached over the environmental focus of Japanese loans, the financing of the loans remains a contentious issue. In contrast to economic infrastructure projects, environmental projects have a low rate of return on investment. In other words, the economic benefits and financial viability of a project to improve atmospheric conditions in Lanzhou are not as obvious in the short term as a new railway line to transport coal from northern China to the southern regions.[69] This factor has important funding implications.

In China, as noted in Chapter 2, local governments are responsible for guaranteeing the repayment of foreign loans. Local governments fund municipal infrastructure projects, while industrial pollution control projects are funded by the industries involved. Foreign/local funding ratios range from between

50:50 to 75:25, depending upon the nature of the project and local currency needs. Local governments often do not have sufficient local currency or the funds to maintain projects and consequently many projects fall behind schedule.[70] The view within SEPA is that the proportion of Japanese funding should be increased to 80 per cent.[71] But the real problem lies with Chinese industry and its reluctance to invest in environmental protection. Unfortunately, the upheaval surrounding the industrial restructuring process in China is making it difficult for enterprises to take on the financial responsibility for environmental loans.

Environmental technology transfer is also an area of continuing donor–recipient tension. The Japanese concept of technology transfer includes both hardware (machinery and equipment) and software (skills and training). The latter conforms to capacity building in the traditional sense of strengthening human resources. Typically, 'technology software' is not fully recognized by developing countries. In fact, confusion often arises over Japan's technical assistance programme, which on balance is more orientated to 'soft' rather than 'hard' technology transfer.

For China, according to the JBIC office in Beijing, 'high technology is the only solution'.[72] But in Japan industrial pollution control technology is owned or patented by private companies, making it difficult to transfer relatively advanced technology on an intergovernmental basis. Environmental technology transfer to China largely takes place through the private sector.[73] On a commercial basis, Japanese technology cannot be licensed in China for ten years after the transfer has taken place, by which time the Chinese argue that the technology will be outdated. It is significant that METI's Green Aid Plan 'does not include technology which is in practical use in Japan'.[74]

Unfortunately, any potential environmental benefits that lower grade technologies may bring have been overshadowed by Chinese suspicions of Japanese intent. In the words of one official at the NDRC: 'the Chinese side does not know the true cost of these environmental technologies and, therefore, must take Japan's word on trust'.[75] Hence, despite Japanese efforts to develop more cost-effective technologies in keeping with the needs of developing countries, actual Japanese environmental technology transfer to China has been limited and largely discounted by the Chinese. Moreover, METI officials admit that it has provided very little follow-on business for Japanese companies.[76]

Environmental aid administration

The earlier tensions are further exacerbated by weak interagency coordination: no single agency in China is responsible for coordinating development assistance. China has a strong preference for keeping foreign donors apart in order to maintain its political control. World Bank loans are negotiated through the MOF, Asian Development Bank loans through the Bank of China and, until recently, Japanese loans were administered by the Ministry of Foreign Trade and Economic Cooperation. The MOF took over the responsibility for Japanese loans in November 1998, which seems to suggest a more cooperative stance on the part of the Chinese government.

DONOR
Policy Agencies

| Ministry of Foreign Affairs | Ministry of Economic Trade and Industry | Ministry of Finance |

Implementation Agencies

| Japan International Cooperation | Japan Bank for International Cooperation |

| Grant aid | Technical cooperation |

RECIPIENT

Policy Agencies

| Ministry of Foreign Trade and Economic Cooperation | Ministry of Science and Technology | Ministry of Finance | National Development and Reform Commission |

Implementation Agencies

| State Environmental Protection Administration |

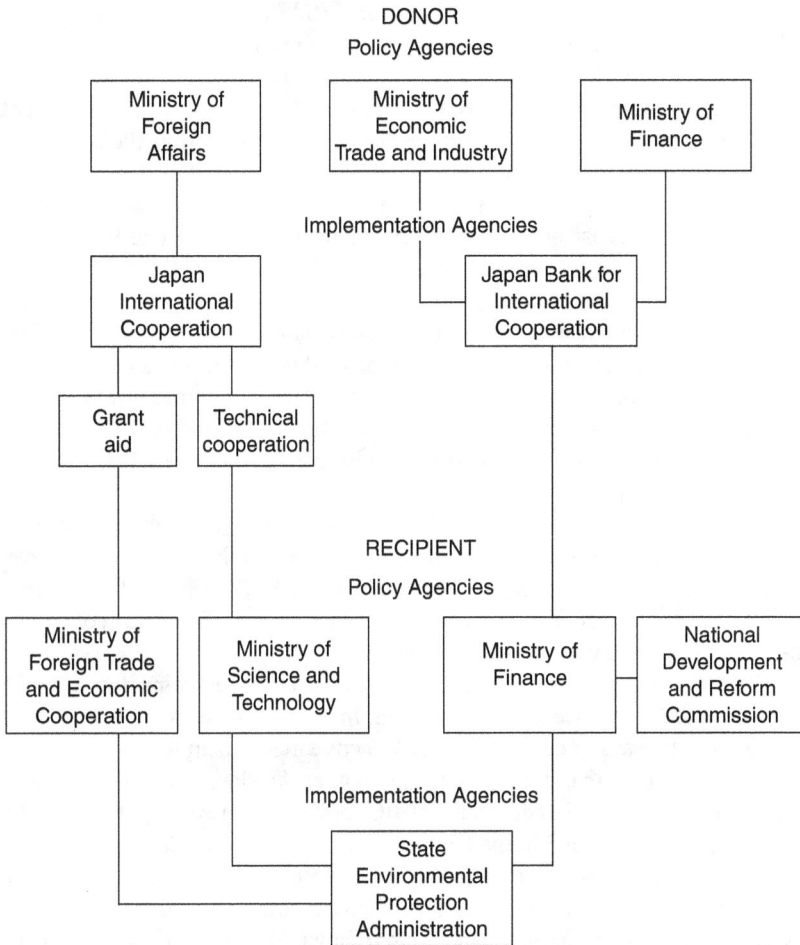

Figure 3.2 The administration of Japanese environmental aid to China.

Source: Compiled on the basis of interviews with Japanese aid agencies.

A total of ten agencies are involved in implementing Japanese environmental aid at the national level in China (see Figure 3.2). Chinese requests for Japanese environmental aid projects are made through three separate agencies – the MOF (loans), the Ministry of Foreign Trade and Economic Cooperation (grants) and the Ministry of Science and Technology (technical assistance). The inclusion of the latter is somewhat of an aberration – at the introduction of Japanese ODA to China in 1979 no Chinese agencies were willing to deal with the Japanese except for the former Science and Technology Commission, many of whose officials were also competent in the Japanese language.[77] These three agencies also consult with SEPA and local governments.

On the Japanese side, actual project selection is supported by the JBIC Special Assistance for Project Formulation Programme (SAPROF): annual missions are sent to China to prepare concrete projects, examine institutional viability and monitor general local capacities. All projects require formal approval from China's State Council and are screened at the JBIC office in Tokyo. The JBIC representative in Beijing is then responsible for coordination at the local level through detailed discussions with provincial governments and city mayors.

From the Japanese perspective, aid implementation at the national level is constrained by China's lack of institutional capacity. According to one JBIC official in Tokyo:

> In general there exists a total lack of coordination between SEPA, the State Planning and Development Commission, local governments, and state firms. Chinese organizations are lacking in initiative when it comes to the environment. The real problem seems to be that nobody in China is willing to bear the final responsibility for environmental projects – not even SEPA![78]

Project staff at JICA are also critical of SEPA's lack of support for technical assistance; for example, when JICA attempted to present the findings of a development study on a wastewater project in Chengdu, SEPA reportedly refused to take an interest.[79] It is significant that no unit exists within SEPA for conducting post-evaluations on environmental projects.

At the same time, China's weak financial system constrains the available options for effective project implementation. In keeping with its own experience, Japan has sought ways to develop a more effective mechanism for funding industrial enterprises in China, but to no avail. In other developing countries such as Thailand, the Philippines and India, JBIC provides a two-step loan facility whereby loans are provided in the first instance to national banks which, in turn, use the loans as a source of funding for private enterprises. In China, the system was tried but failed as a result of structural weaknesses within the Chinese banking sector. Consequently, Japanese environmental loans can only be directed to SOEs. Given the huge demand for environmental protection within the Chinese state-owned sector, the important issue has become 'how to use financial transfers as pump-priming funds for the greatest effect'.[80]

In addition to the above institutional and financial constraints, a major weakness in the aid agenda setting process appears to be the limited room for involving local stakeholders. Local environmental needs and priorities are factored into the negotiation process indirectly. Provincial and municipal EPBs may propose requests for international environmental assistance through SEPA, but these requests are only met if they accord with state plans for environmental protection. The response of the local EPBs to their exclusion is mixed: some feel frustrated that national priorities are weakening the local capacity for comprehensive environmental protection; others are more modest in recognition of the fact that they lack both the necessary experience and knowledge to negotiate effectively with international donors.[81] Either way, it would seem that international donors have an important mediating role to play in reconciling central government interests with local government needs and priorities. In the Japanese case, this may be facilitated by a new method of aid disbursement introduced in October 2001. Instead of the usual multi-year programmes, Japanese aid is now disbursed on an annual project-by-project basis. This means that Japanese aid officials are now more actively involved in screening projects and working jointly with central and local governments.

Implementing environmental projects at the local level

The six selected project sites in this case study (see Map 3.1) include both the first round of environmental improvement projects introduced in 1996

Map 3.1 Regional location of selected Japanese environmental projects.

Source: National Geographic, *Atlas of the World*, 5th edn, Washington, DC: National Geographic Society, 1981.

(Shenyang, Benxi, Huhhot and Liuzhou) and the model city projects which began implementation in 1999 (Dalian and Guiyang). To provide comparative analysis, three of the projects are located in the relatively developed region of Liaoning province and the other three in the poorer regions of Inner Mongolia Autonomous Region, Guizhou province and Guangxi Zhuang Autonomous Region. Liaoning province has made some progress in its transition towards a market economy, whereas in the north-central and southwestern regions of China the local economies are still firmly under administrative control.[82]

Japanese environmental improvement projects

The Japanese environmental improvement projects are currently at different stages of implementation and are due for completion by 2006. The key environmental objectives for the projects include:

1   Abatement/treatment of industrial pollution (gaseous emissions, dust, wastewater, liquid effluents and solid wastes).
2   Development/expansion of public utilities (gas supply, central heating system, water supply, sewage treatment and garbage management).
3   Promotion of environmental protection industries and institution building for environmental monitoring and technological and human resource management.[83]

In practice, the projects are heavily biased towards the first objective, with some emphasis upon the second but very little evidence of the third.

The specific environmental targets for each project are in keeping with the top pollution abatement priorities for each city – TSP and SO_2 in Shenyang, heavy metal discharges in Benxi, TSP and dust in Huhhot and SO_2 and acid rain in Liuzhou. It is significant that the Japanese feasibility study for the projects carried out under the auspices of JBIC was little different from the feasibility studies prepared by the municipal EPBs themselves.[84] Clearly, local EPBs do have the necessary competence to identify the major sources of environmental problems and systematically work out solutions, albeit in a narrow technocratic way. In some cases, the task was made considerably easier by the fact that the most severe industrial pollution was highly centralized amongst one or two enterprises, as in the case of Shenyang. But in other cities, as in the case of Benxi, the local feasibility studies involved more than 15 different enterprises.

Project implementation is the responsibility of the enterprises involved under the supervision of the municipal EPBs. Leading groups have been established to oversee the project implementation process which are headed by the city mayors and include high-ranking local officials from key agencies such as the Planning Committee, Finance Bureau, Construction Bureau, Economic Trade Committee and the Bureau of Public Affairs.[85] It is, however, difficult to gauge the extent to which the leading groups actually serve a purpose beyond their official stamp of approval. The day to day supervision and monitoring of the projects is carried out by the project implementation unit within the municipal EPB and relies upon 5–6 dedicated staff. The exception to the rule is in the city of Shenyang where the project implementation unit resides in the Shenyang Planning Committee. All four projects are, therefore, similar in their environmental management focus and local institutional arrangements. The difference lies in the local social and economic conditions under which the projects are implemented, especially in relation to the degree of industrial restructuring.

Shenyang environmental improvement project

Shenyang (population 6.71 million) is the provincial capital of Liaoning province, which is home to China's oldest industrial base.[86] Over the past decade, efforts to restructure the regional economy have met with some success, although this has carried high social costs. According to Louise Cadieux, between 1994 and 1997, 1.2 million workers were laid off – the highest number of any province in China.[87] Historically, Liaoning's large industrial sector has placed a heavy burden on the environment. For example, in 1985 the annual average indexes for TSP and SO_2 in Shenyang were over 500 $\mu g/m^3$ in comparison to WHO standards of 60–90 (TSP) and 40–60 (SO_2).[88] The chemical oxygen demand (COD) in the two main rivers passing through the city (Liao and Hun) was below China's lowest national standard for most of the year, causing the rivers to turn black. By 1988, Shenyang had become one of the ten most polluted cities in the world. Ten years later, following the introduction of a number of industrial reforms, TSP and SO_2 levels declined to 280 and 97 $\mu g/m^3$ respectively, but still exceeded the national standards by 80 per cent; the city had no sewage treatment plant and over 50 per cent of urban drinking water did not meet national standards.[89]

Given the history and severity of urban pollution problems in Shenyang, it is not surprising that the city has become a top priority for international environmental assistance. Shenyang is the only city in China that has received assistance for urban environmental management from all of the leading donor organizations – World Bank, JBIC, UNDP and the European Union.[90] JBIC is by no means alone in providing environmental assistance to Shenyang, although it was one of the first donor organizations to stand in line.

Project discussions between the central government and JBIC started in 1990 and lasted six years before a loan package of US$54 million was finally disbursed (see Table 3.2). The aggregate cost of the project totalled US$100 million with the remainder provided by the municipal government and the Chinese enterprises involved. A Green Project Office was set up under the supervision of the Shenyang Planning Committee rather than the EPB (which reflected the relative power of the Planning Committee *vis-à-vis* the EPB at the time).[91]

The Shenyang project has focused largely upon reducing SO_2 and TSP emissions from the Shenyang Smelter, which in 1999 was a leading source of air pollution within the municipality – contributing 42 per cent of SO_2. At the commencement of the project, the average density of SO_2 in the central district of the Shenyang Smelter was 190 $\mu g/m^3$.[92] Equipment and technology for the project

Table 3.2 Shenyang environmental improvement project 1996–2003

Sub-project enterprise/ institute	Project description	Target pollutants
1996 projects		
Shenyang Smelter	Renovation of copper smelting and acid making system	TSP, wastewater, SO_2
Shenyang Power Station (3 × 200 MW)	Renovation of boiler system	Dust, coal, SO_2
1999 projects		
Shenyang Alloy Company	Renovation of smelting system	Dust, acid mist, SO_2
Shenyang Heat Supply Company[a]	Centralization of power stations for heat supply	Dust, coal, SO_2
Shenyang Jinshan Thermal Power Station	Installation of circulating fluidized boiler system	TSP, SO_2
2001 projects		
Shenyang Thermal Power Station	Installation of thermoelectric power supply	TSP, SO_2
Industrial plants	Renovation of boiler and smelting systems	TSP, SO_2

Source: Compiled from OECF, Special Assistance for Project Formation Study, 'People's Republic of China environmental improvement project – interim report of China environmental improvement project', unpublished, Tokyo: OECF, March 1995; SEPA, 'Riyuan daikuan huanbao xiangmu jishu yu xinxi jiaoliu zhinan' (Japanese yen loan environmental protection projects: technical and information guide), unpublished, Beijing: SEPA, 1999; JBIC, *JBIC Annual Report*, Tokyo: JBIC, 2002; JBIC, *JBIC Annual Report*, Tokyo: JBIC, 2003.

Note
a Public utility project.

have been, for the most part, procured locally with some sourcing from overseas – not including Japan. The project has often stalled as a result of insufficient local funding. But the underlying problem seems to be one of ownership; interviews outside of Shenyang revealed that the Smelter and the Shenyang Power Station were owned by central government officials in Beijing who had reportedly been unwilling to forfeit enterprise profits for the sake of municipal environmental improvement.[93]

Officials at the project office in Shenyang were mainly concerned with the cost-recovery problems involved with air pollution projects and, therefore, suggested that Japanese environmental loans should be directed towards water pollution projects. Project officials were of the opinion that project implementation was progressing more smoothly in Benxi and Dalian because 'they have water projects which are easier to operationalize and evaluate'.[94] It was conceded, however, that with appropriate pricing incentives the cost-recovery problem could be alleviated. The Jinshan Power Plant in Shenyang, for example, was in a position to recoup investment in air pollution control by increasing electricity charges.[95] In addition, the Shenyang office stressed the need for more comprehensive project feasibility studies. In the words of the Director:

> In general the responsibility of the project office is very difficult. Loan agreements are signed but not enough attention is given to pre-feasibility studies to ensure that projects can actually be implemented. It is not possible for us to do this ourselves because we already have to prepare many documents for different agencies amongst the Chinese government in addition to the OECF.[96]

Despite these concerns, between 1996 and 1999, as a result of Japanese financial assistance, the Shenyang Smelter reduced its SO_2 emissions by 12 per cent.[97] In 1999 three additional projects were initiated. These involved air pollution control in two enterprises together with a public utility project to support Shenyang's efforts in the centralization of heat supply. At the time of my visit to Shenyang in May 1999, these new projects were still at an early stage of implementation, however local officials seemed fairly confident that the Shenyang Smelter project was leading to significant improvements in pollution abatement. In this particular case, the outcome will never be known. In January 2001, the Smelter was closed down by the central authorities in Beijing. This action was viewed by many as a symbolic sign of the Chinese government's commitment to environmental reform.[98] But for donors it highlighted the serious risk that they face in wasting millions of dollars in controlling pollution from failed SOEs. Two months later, on 30 March 2001, Japan provided a further loan of ¥6,196 million (US$50.6 million) for the purposes of improving air quality in Shenyang. This time the project was framed more broadly on the basis of anti-global warming.[99]

Benxi environmental improvement project

Benxi (population 1.5 million) is situated in the eastern mountainous region of Liaoning province. The city is an important industrial raw material base, producing predominately iron, steel and cement. But for decades it has also produced tons of pollution and waste, made worse by the city's location in a valley, which reduces the potential for the natural dispersion of low altitude emissions. In 1980, air pollution was so severe that the city was no longer visible by satellite. Between 1970 and 1982 the death rate from cancer rose from 113 per million to 369 per million – an increase of 230 per cent! As in the case of Shenyang, TSP and SO_2 levels were exceptionally high at around 500 µg/m³.[100] Significant pollution control did not occur until Benxi was listed as a pilot city for national environmental protection control in 1989. Under a seven-year programme to control 22 out of the 33 most serious 'dust dragons' (dust-emitting stacks), dust levels dropped from 53.2 tons per km² on a monthly basis in 1988 to 39.9 tons per km² in 1996.[101] Immediately following the programme, JBIC provided a loan for US$100 million which largely focused upon industrial water pollution control including the construction of a water treatment plant and a municipal gasification facility (see Table 3.3).

Project negotiations began in 1994. Japan had a special interest in Benxi because Japanese companies had been conducting environmental protection business in the city since 1989. Initially, the project involved 39 sub-projects but because of changes in enterprise ownership – many enterprises were sold to district governments – the total number was finally reduced to 18 sub-projects. Unfortunately, delays in project implementation have meant that the overall environmental effects of the project are still not clear.

The focus upon water pollution as opposed to air pollution reflects local enterprise concerns over the cost of air pollution abatement.[102] But at the time of my visit to the project in 1999, both water and air pollution projects were experiencing financial difficulties: the enterprises involved had either been unwilling or unable to raise the necessary local currency (typically 50 per cent of the total value of the project). Many enterprises in Benxi were facing increasing pressure to restructure and reform, especially following the loss of export markets in Southeast Asia as a consequence of the 1997 Asia financial crisis.

The low interest rates offered by Japanese environmental loans were providing an important incentive for these industries but this was not sufficient to allay local concerns.[103] The project office in Benxi maintained that the time lag involved between the application and actual disbursement of environmental loans for pollution control was ill suited to the current pace of industrial reform. One EPB official pointed out that those enterprises applying for an environmental loan in 1999 would have to wait until 2002 for disbursement by which time the enterprise could be facing closure.[104] This again reinforces the need for more comprehensive project feasibility studies. In addition, according to the Director of the project office, the experience of implementing Japanese environmental loans had increased local understanding of the need for a stronger institutional environment:

> The local government must provide more local currency and reduce tax levies for environmental protection investment. Provincial and municipal finance bureaux should guarantee the loans. Without such a guarantee we are always concerned that if there is a problem with project implementation the provincial government will withdraw funding from the project.[105]

Huhhot environmental improvement project

Huhhot (population 938,000) is the capital city of the Inner Mongolian Autonomous Region that stretches 4,221 km across China's northern territory bordering Qinghai province in the west and Heilongjiang province in the northeast. Much of the region comprises vast areas of grasslands (880,000 km^2) which over the years have become heavily degraded as a consequence of over-grazing. Pollution problems in the region are equally serious.[106] Huhhot experiences

Table 3.3 Benxi environmental improvement project 1996–2003

Sub-project enterprise/institute	Project description	Target pollutants
1996 projects		
Benxi Iron and Steel Company	Installation of wastewater treatment plant	S_2, COD
Benxi Chemical Plant	Installation of wastewater treatment plant	S_2, COD
Benxi Alloy Company	Installation of gas treatment and water treatment facilities	Cu, Benzene
Benxi City Electroplating Plant	Relocation and wastewater treatment	Wastewater (especially heavy metal Cn)
Benxi Bearing Plant	Installation of gas treatment and water treatment facilities	Dust, wastewater
Benxi Pharmaceutical Plant	Wastewater treatment	BOD
Benxi Electric Appliance Plant	Relocation	Dust, Cn
Benxi Iron and Steel Company	Desulphurization and dust collection	SO_2, TSP
Benxi Alloy Company	Installation of wastewater and effluent treatment	Gas, acid mist, NH_3, SO_2
Benxi Cement Company	Dust collection	Dust
Gongyuan Cement Company	Dust collection	Dust
Xiaoshi Cement Factory	Installation of electrostatic dust collector	Gas, dust
1999 projects		
Beitai Iron and Steel Company	Installation of dust removal system, gas purification and sewage treatment system	Gas, TSP, wastewater, SO_2
Benxi Rubber Chemical Plant	Relocation	NO_x, wastewater, SO_2
Benxi Lubricant Factory	Wastewater and gas treatment	Wastewater, H_2S, SO_2
Benxi Auxiliary Agent Plant	Renovation of production system	Dust, wastewater
Benxi Water General Corporation[a]	Construction of water treatment plant	Municipal wastewater
Benxi Gas General Company[a]	Construction of city gasification facility	Coal
2003 projects		
Beitai Iron and Steel Company	Wastewater treatment	Wastewater
Industrial enterprises	Desulphurization equipment for coke ovens	NO_x, SO_2

Source: Compiled from OECF, Special Assistance for Project Formation Study, 'People's Republic of China'; SEPA, 'Riyuan daikuan huanbao'; and JBIC, *JBIC Annual Report*, 2003.

Note
a Public utility projects.
Chemical symbols: S_2 (sulphur), COD (chemical oxygen demand), Cu (copper), Cn (Cyanide), BOD (biological oxygen demand), NH_3 (ammonia), H_2S (hydrogen sulphide), NO_x (nitrogen oxide).

China's highest levels of TSP, especially during the heating season (for 98 days of the year the temperature is below $-10°C$) when average concentrations reach between 700 and 800 $\mu g/m^3$ (exceeding WHO standards by 10 times). Industry is the main source of airborne particulates (45 per cent) followed by natural dust from the Gobi desert (30 per cent) and households (25 per cent). Water pollution is also serious – water resources are scarce and consequently, the groundwater has become polluted as a result of over-extraction. All sewage and industrial waste is discharged untreated into the three rivers of the Da Xie, Xiao Xie and Xi.[107]

Unlike the situation in Liaoning, Inner Mongolia has received limited foreign assistance for its environmental problems. The World Bank has invested in a wind power project in Huitangxilot and both the Australian and German governments have provided grant assistance for grasslands conservation. A lack of international funds is evident from the relatively poor conditions of the municipal EPB. In comparison to its counterparts in Shenyang and Benxi, the Huhhot EPB is located in an old building and has fewer and less well educated staff.

In 1996, Japan was the first donor to provide assistance for urban environmental management. The Japanese environmental project in Huhhot (total US$54 million) comprises eight sub-projects mainly related to TSP control (see Table 3.4).[108] The focus on air pollution control is in keeping with the priorities of the central government – bearing in mind that TSP control in Inner Mongolia also relieves dust levels downwind in Beijing. It also remains a local priority: at public hearings involving 10,000 citizens in 1992 and 1993, 80 per cent supported immediate action for improving air pollution, in particular dust.

Procurement for equipment and technologies for the project was carried out under a system of open international bidding; no Japanese technologies were used because they were considered to be too expensive. According to the Director of the Huhhot EPB, local control over funding was having a positive effect on financial capacity because the EPB had been able to expand its knowledge of the most cost-effective and appropriate environmental technologies that were available on the international market. Furthermore, unlike the situation in Benxi, interviews at both the regional and municipal EPBs revealed that industry compliance was not causing a major problem for project implementation. Dust control is relatively inexpensive compared to desulphurization, hence the need for a return on investment is less crucial. Moreover, industry was not under immediate pressure from the central government to restructure and improve market efficiency because Huhhot's overall level of economic development was still relatively low.[109]

In general, project officials responded more positively to Japanese environmental assistance than their counterparts in Shenyang and Benxi. They claimed that the Japanese environmental loan projects had increased local political support for environmental protection and in turn raised the status of the local EPB *vis-à-vis* local planning agencies. The local mayor was keen to encourage international support because local funding for environmental protection was not readily available. The overall environmental impact of these projects has yet to be determined, which suggests weak monitoring capacity on the part of the local EPB.

Table 3.4 Huhhot environmental improvement project 1996–2003

Sub-project enterprise/institute	Project description	Target pollutants
1996 projects		
Heating Power Company of Huhhot City[a]	Expansion of heating supply	TSP
Huhhot Gas Corporation[a]	Installation of gas tank and pipeline	TSP
Huhhot Chemical Plant	Dust reduction	TSP
Huhhot Iron and Steel Plant	Installation of dust collection system and water treatment plant	TSP, SO_2
Huhhot Iron Works of Inner Mongolia	Installation of new equipment to reduce carbon	CO, save coal
Inner Mongolia Chemical Fibre Plant	Wastewater treatment	CCl_4 gas, COD, S_2, Zn
Huhhot Rubber Plant of Inner Mongolia	Renovation of heat and power supply systems	Dust, SO_2
Huhhot Sugar Refinery	Installation of pollution treatment	Smoke, slag
2003 projects		
Huhhot City Sewerage Company	Construction of four sewage treatment plants, sewerage pipes, training for employees	Wastewater

Source: Compiled from OECF, Special Assistance for Project Formation Study, 'People's Republic of China'; and JBIC, *JBIC Annual Report*, 2003.

Note
a Public utility projects.
Chemical symbols: CO (carbon monoxide), CCl_4 (carbon tetrachloride), COD (carbon oxygen demand), S_2 (sulphur), SO_2 (Sulphur dioxide), Zn (zinc).

However, the local support for Japanese environmental loans has paid off. As part of the 2003 ODA package to China, the Huhhot municipal government received a further ¥9.7 billion to improve water quality and sewage treatment.

The rise and fall of the Yellow Dragon in Liuzhou

Liuzhou (population 780,000) is situated in the middle reaches of the Liu River in Guangxi Zhuang Autonomous Region in southwest China. It is now one of China's worst acid rain areas, largely because of the high sulphur content (5–9 per cent) of local coal. The frequency of acid rainfall is around 90 per cent with a pH value below 4. Industry is the main source of SO_2 emissions (81 per cent) with approximately 80,000 tons of SO_2 discharged into the atmosphere annually.[110] Liuzhou municipal government started a SO_2 abatement programme in the mid-1980s with limited success. In 1990, the central government listed Liuzhou as one of the trial cities in China for implementing SO_2 discharge permits. Total emissions of SO_2 (annual average concentration) declined from 217 $\mu g/m^3$ in 1993 to 164 $\mu g/m^3$ in 1997 (still far exceeding the national average of 79 $\mu g/m^3$).[111]

Table 3.5 Liuzhou environmental improvement project 1996–2003

Sub-project enterprise/ institute	Project description	Target pollutants
1996 projects		
Liuzhou Heating Power Company[a]	Central heating-power project and reform of boiler system	SO_2, dust, coal
Liuzhou Maintenance and Administration Department[a]	Liuzhou-Longquanshan Sewage Treatment Plant	Municipal wastewater
Liuzhou Dongfeng Chemical Plant	Installation of desulphurizing agent and construction of slag cement plant	SO_2
Liuzhou Iron and Steel Company	Coke oven gas desulphurization	SO_2
Liuzhou Zinc Production Company	Relocation of plant facilities to 13 km outside of the city centre	SO_2, wastewater
Liuzhou Fertilizer Plant	Treatment of tail-gas from nitric acid	NO_x
1999 projects		
Liuzhou Power Plant (2 × 200 MW)	Installation of flue gas desulphurization process	SO_2

Source: Compiled from OECF, Special Assistance for Project Formation Study, 'People's Republic of China'; and SEPA, 'Riyuan daikuan huanbao'.

Note
a Public utility projects.

Japan was the first country to provide environmental assistance to Guangxi, beginning with a JICA funded development study for integrated air quality management in Liuzhou, Guilin and Wuzhou. In 1994, JICA established ten monitoring stations to check daily air quality. The JBIC environmental loan in 1996 involved 2 public utility projects and 5 industrial pollution control projects at a total cost of US$100 million (Table 3.5). The 1996 projects have sourced technologies and equipment locally, with some enterprises using patented technologies from Tsinghua University in Beijing. As in the case of the Japanese project in Huhhot, local control over funding has stimulated a strong interest in seeking out the most appropriate and cost-effective environmental technologies.[112]

At the time of my visit, some of the sub-projects had already provided some environmental benefits. At the Liuzhou Iron and Steel Plant, SO_2 emissions had decreased by 249 kg per hour with the expectation that a further 50 per cent reduction would be made upon completion of the project.[113] Liuzhou's *Huang Long* (Yellow Dragon) (referred to in the introduction), had not been slain but considerable progress had been made – the tail gas of nitric acid had decreased from 3,500 parts per million (ppm) to 1,000 ppm with a further reduction of 500 ppm expected at the project's end.[114] The environmental effects of the project were relatively easy to verify because at the commencement of the project JBIC had provided funds for an acid rain monitoring centre.

As in the case of Huhhot, EPB officials responsible for the overall implementation of the project were not experiencing any problems with industry compliance. The local economy was still at a relatively low level of development and industrial restructuring and SOEs reform lagged far behind Liaoning province. One EPB official also made the point that the Liuzhou Iron and Steel Plant and Liuzhou Power Plant sub-projects were the same as those in Shenyang but implementation was smoother because the factories were strictly under local government control.[115]

The Liuzhou project office was particularly supportive of Japanese environmental assistance, which is perhaps a reflection of the strong mutual interests involved in reducing acid rain. The Director of the Liuzhou EPB described Japanese environmental loans as 'efficient, with simple operating procedures and good results'.[116] The degree of political commitment was evident from the fact that Liuzhou's leading group for acid rain, headed by the mayor, had been established in 1993 before the commencement of the project. This was not the case at other project sites.

Japanese environmental model city projects

The purpose of the environmental model city projects is to improve atmospheric pollution in a more comprehensive manner by promoting energy efficiency, industrial recycling, environmental monitoring and technical training. A secondary objective is to encourage the local production of low-cost pollution equipment to meet China's specific needs. The model city concept is designed not only to concentrate resources but also to demonstrate practices that could be replicated nationally. In the words of Toshio Watanabe head of the Japanese expert committee for the projects:

> The Japan–China environmental model city initiative is aimed at creating the impetus for China to act on its own to protect the environment. Its aim is not just to fight individual sources of pollution but to create a mechanism whereby environmental protection efforts will be expanded outside the model cities.[117]

Three Chinese cities – Dalian, Chongqing and Guiyang – were chosen in order to promote environmental protection at two different levels: to increase air pollution control in Chongqing and Guiyang to a level comparable with Dalian, and to improve comprehensive environmental planning in Dalian to a level comparable with Tokyo.[118] JBIC provided financial support for the model city concept in the last two years of the fourth batch of ODA loans (1999–2000) at a total cost of approximately US$1.5 billion divided between the three cities.

Dalian environmental model city project

Dalian (population 5.3 million), situated on the southern tip of the Liaodong Peninsula, is the most economically advanced city in northeastern China.

However, long-term economic development is handicapped by a large number of loss-making SOEs – steelworks, cement, chemicals, textiles and pharmaceuticals – that are currently in the process of market reform. Burdened with inefficient and outdated equipment and technologies these industries are largely responsible for the city's smoke and soot problems and its heavy coastal water pollution. The basic environmental infrastructure in Dalian is relatively good, but the city still has a low rate of industrial recycling, and SO_2 and TSP emissions exceed WHO standards by a factor of three (in 1996, SO_2 and TSP concentrations were 137 and 264 $\mu g/m^3$ respectively). The city is also experiencing problems with NO_x emissions from automobiles (88 $\mu g/m^3$ in 1996), although emissions are low in comparison to Guangzhou and Shanghai.[119]

Japanese environmental assistance began in the late 1980s with local government cooperation between Dalian and Kitakyûshû, the leading industrial city of southwestern Japan. The Kitakyûshû government in 1990 first proposed the environmental model city concept. The term is misleading; what is actually meant is a model district or zone within the city. JICA provided the technical assistance for a research and development study at a cost of US$100 million for three consecutive years 1996–1999 with the intention of attracting OECF/JBIC financial support. The study involved detailed plans for restructuring industry, improving energy efficiency, introducing cleaner production processes and strengthening monitoring and regulatory supervision. A strong emphasis was also placed upon the need to establish an environmental industrial organization for private companies, an environmental protection fund and a citizens environmental association.[120]

It is important to mention here that although the Kitakyûshû–Dalian relationship provides a useful channel for Japanese environmental assistance, in practice building good relations at the local level have proven difficult. The Dalian government prefers to keep local and national cooperation apart. In the words of the JICA expert responsible for the implementation of the development study: 'the Dalian government has two tongues – one to speak to the Kitakyûshû government and one to speak to JICA'.[121] Moreover, local government relations between Japan and China do not appear to have affected the scope of JBIC's environmental assistance to any significant degree. The environmental model city project in Dalian currently differs little (if at all) from the environmental improvement projects described earlier.

The JBIC environmental loan provided in 1999 includes six sub-projects involving municipal wastewater and sewage treatment and industrial pollution control in 2 factories and 2 power stations (see Table 3.6). At the time of my investigation, the project was still at an early stage of implementation but already a major problem was apparent. The two factories involved were both heavy polluters and were operating at a substantial loss. In the context of Dalian's economic reforms, there existed a high risk that the Japanese loans would help to prolong the lives of dirty factories that would otherwise have been closed down on economic and environmental grounds.

The problem can best be illustrated by the case of the Dalian Cement Factory. This factory has been in operation since 1907 (then under the ownership of a Japanese company, Onada) and most of the equipment in current use dates back

Table 3.6 Dalian environmental model city project 1999–2003

Sub-project enterprise/ institute	Project description	Target pollutants
1999 projects		
Wafangdian Municipal Tap-water Company[a]	Construction of water treatment plant	Municipal wastewater
Wafangdian Management Bureau for Urban and Rural Construction[a]	Construction of sewage treatment plant	Municipal sewage
Dalian Pharmaceutical Company	Relocation and installation of water treatment facilities	Wastewater
Dalian Cement Factory	Installation of dust collectors	Dust, TSP
Dalian Chunhai Thermal Power Plant	Centralization of boiler facility	TSP, SO_2
Dalian Yandao Thermal Power Plant	Centralization of boiler facility	TSP, SO_2
2001 projects		
Thermal Power Plants	Construction of energy efficient boilers and generating units	TSP, SO_2
Industrial enterprises	Installation of water treatment and dust collectors	TSP, SO_2

Source: Compiled from SEPA, 'Riyuan daikuan huanbao'; and JBIC, *JBIC Annual Report*, 2003.

Note
a Public utility projects.

to the early 1950s. In 1999, it was running at a reduced annual capacity of 550,000 tons in comparison to the annual average capacity of 730,000 tons. The factory had no electrostatic dust collection facilities and therefore was producing serious fugitive dust (it is not insignificant that the factory is located next to the Dalian international airport). In response, the local government had closed down two of the cement mixers causing the worst pollution. The factory had also been ordered by the municipal government to meet national air pollution standards by 2002. The question still remains as to whether Japanese environmental assistance will resolve or merely prolong the factory's predicament.[122]

Dalian EPB officials were of the opinion that JBIC assistance was not suitable because it only provided investment for equipment and not for the operation and maintenance of projects which was essential to alleviate cost-recovery problems. They maintained that Japanese environmental loans could have little impact without the necessary technical assistance. In particular, Dalian EPB officials were critical of the fact that Japanese loans did not provide assistance for cleaner production, which was seen as crucial in order to convince local industries of both the environmental *and* economic benefits of pollution control.[123] Some industries in Dalian were already convinced, as claimed by the vice-director of the Dalian Cement Factory: 'We need cleaner production to increase process efficiency and the efficient utilization of raw materials.'[124]

Table 3.7 Guiyang environmental model city project 1999–2003

Sub-project enterprise/ institute	Project description	Target pollutants
1999 projects		
Guiyang City Gas Company[a]	Extension of municipal gas supply	Coal
Guiyang Steelworks	Installation of dust collection facility, desulphurization	Dust, SO_2
Guizhou Electric Power Plant	Flue gas treatment and renovation	Dust, SO_2
Guiyang Hongyan Chemical Plant	Relocation and wastewater treatment	Wastewater
Guizhou Cement Plant	Installation of electrostatic dust collectors and relocation	Dust
Lingdong Coal Administration	Cleaner coal processing	SO_2
Guiyang Environmental Monitoring Station	On-line monitoring systems for air quality and pollutant sources	N/A
2001 projects		
Guiyang City Gas Company	Construction of gas supply facilities	Coal
Industrial enterprises	Installation of desulphurization equipment	Dust, SO_2

Source: Compiled from SEPA, 'Riyuan daikuan huanbao'; and JBIC, *JBIC Annual Report*, 2003.

Note
a Public utility project.

Guiyang environmental model city project

Guiyang, with a population of 3.5 million, is the capital of Guizhou province in southwest China. A heavy reliance upon low grade coal (with a sulphur and ash content of over 5 per cent) means that the province suffers from a high frequency of acid rain with a pH value of between 3 and 4.5. The impact upon Guiyang's environment is so severe that zinc-plated steel lasts only 4 years in contrast to the usual 29 years. Historically, the city has experienced the highest levels of SO_2 in China, reaching 407 $\mu g/m^3$ in 1985 – 10 times the WHO standard of 40–60 $\mu g/m^3$.

The municipal government has made some effort to control SO_2 emissions through regulatory enforcement and industrial relocation – SO_2 emissions decreased to 300 $\mu g/m^3$ in 1996. But the revenue gained from pollution discharge fees is only a fraction of the cost of pollution control and, as a consequence of the low level of economic development, the Guiyang EPB has experienced difficulties in accessing local government funds for environmental improvement.[125] The total amount of investment resources for pollution treatment in Guizhou is only one-fifth of the equivalent resources in Liaoning and less than 50 per cent of resources in Inner Mongolia.[126]

International environmental assistance to Guizhou is also limited to Japan and the UNDP. The JBIC environmental city project involves seven sub-projects largely related to SO_2 abatement and includes an environmental monitoring station (see Table 3.7). Most of the enterprises involved have sourced the necessary technologies either locally or through a system of international bidding. The one exception is the Guiyang Power Station that has imported technologies from Japan. It has used Japanese desulphurization technology with an efficiency rate of 90 per cent to reduce SO_2 emissions from 87 million tons per annum to 17 million tons.[127]

As in the cases of Huhhot and Liuzhou, local control over funding has stimulated an interest in cost-effective environmental technologies. Moreover, without the immediate political pressure to reform SOEs, EPB officials claim that they have not experienced any problems with industrial non-compliance. On the contrary, according to the Director of the project office, Japanese loans have provided local factories with an incentive to take environmental protection more seriously.

Strong ownership or weak capacity?

The above project investigations have demonstrated that Japanese environmental assistance makes an important financial contribution towards improving environmental infrastructure and point-source pollution control in China. The fact that urban industrial pollution in Chinese cities is concentrated amongst a few major industrial sources allows the Japanese to target the most seriously polluting factories in each city. Given the size of these factories, the potential environmental impact is quite considerable and already discernible in the cities of Shenyang and Liuzhou.

In general, the implementation of Japanese environmental loan projects has been smoother in the poorer southwest and north-central regions of China than in the more developed northeast. In part, this is because the demand for large-scale environmental infrastructure is far greater in the poorer regions where local funds are less readily available. In poorer regions external funding can also have a powerful effect upon local political commitment. It can encourage the support of senior government officials and, in turn, raise the political status of the local EPB. As witnessed in the enthusiastic behaviour of the mayor of Liuzhou, sometimes the social incentives relating to pride and local reputation are as important as financial incentives in enhancing political commitment.

Above all, the relative effectiveness of the Japanese approach to environmental management seems to depend upon *the degree of industrial restructuring* within the regional setting. In the poorer, regulated local economies of Huhhot, Guiyang and Liuzhou where industrial restructuring lags far behind, local governments can focus their attention solely upon pollution control. Japanese environmental assistance provides much needed financial investment for those industries that are likely to remain the backbone of the local economy in the short to medium term. Local perceptions based upon interviews with the project implementing agencies support this conclusion: concerns over economic viability and a financial return on investment were far higher in Liaoning province than in Inner Mongolia, Guangxi or Guizhou.

Conversely, in the mixed local economies of Shenyang, Benxi and Dalian, where industrial restructuring is relatively advanced, local governments are under considerable pressure to deal with industrial reform and pollution control simultaneously. Hence Japanese point-source control projects run the risk of diverting much needed resources into loss-making factories that are likely to be closed down in the immediate future. This is because Japanese environmental assistance currently targets the most polluting factories in any given city regardless of the economic feasibility of doing so. One of the consequences in the northeast of China has been a de facto subsidization of non-performing industries, which may well produce the undesirable effect of stalling the reform process and thus be counterproductive in regard to reducing emissions in the long term.

In evaluating the effectiveness of the Japanese engineering approach in terms of the four broad dimensions of environmental capacity – financial, technological, institutional and social – a different picture emerges that reveals longer term problems regardless of the regional setting. Japanese environmental assistance has not contributed in a significant way to building environmental capacity. Japan's greatest strength in environmental management lies in its experience in technological innovation. Yet Japanese environmental loans are having a minimal effect upon China's technological capacity for environmental improvement; they are not being used to implement cleaner production which local officials themselves seem to have a strong interest in promoting.

It is interesting that, in stark contrast to priorities at the national level, the issue of advanced technology transfer at the local level is not so important. Most enterprises involved in Japanese environmental loan projects believed that they could source the necessary technologies domestically, and that importing state-of-the-art environmental technologies, especially from Japan, was not cost-effective. The Liuzhou Chemical Fertilizer Plant, for example, was using a method borrowed from the Nanjing Chemical Industry Company to treat NO_x emissions with the by-products being used for fertilizer, thus reducing costs at the same time. The experience gained was encouraging the senior management of the enterprise to make further investments in environmental protection.

On a more optimistic note, the Japanese approach does seem to have had a positive effect on local financial capacity simply by virtue of the fact that Japan does not impose its own environmental agenda or restrictive loan conditions. Local beneficiaries enjoy a strong sense of ownership over financial resources, and this, in turn, has implications for their capacity to utilize funding effectively. In general, the efforts made by local EPBs to access information worldwide and assist local enterprises in selecting the most appropriate technologies and processes for industrial pollution control have been quite impressive. Likewise local enterprises, particularly those faced with the high costs of desulphurization in the acid rain areas of Guiyang and Liuzhou, have made serious efforts to secure the most cost-effective means of utilizing Japanese environmental investment.

In effect, the advantage of Japanese environmental assistance lies not in the transfer of environmental knowledge and technologies, but rather in the transfer of financial resources to enable local officials and enterprise managers to pursue local environmental goals as they see fit. As noted at the beginning of this chapter, 'ownership' is one of the key principles guiding Japan's efforts in international environmental assistance. The problem is that strong ownership, or more specifically local control over financial resources, does not always equate with capacity building. Ownership over financial resources does not seem to have had much effect upon financial capacity in China's more developed regions where the economic viability of environmental investment is a major issue because of the ongoing economic reforms. Moreover, the Japanese approach may enhance the effective utilization of funding in the short term but it provides no guarantee of recurrent funding to cover operational and maintenance expenditure.

If Japan is serious about improving the effectiveness of its environmental aid then it needs to focus upon building local capacity by promoting cleaner production, economic incentives and public participation. In so doing, the emphasis upon ownership needs to move beyond a small group of beneficiaries to a broader public involvement that better reflects the economic and social interests of the local community at large. This would have the added bonus of alleviating Japanese public concerns that their aid is not fully appreciated in China.

Conclusion

This case study has shown that a traditional technocentric approach to environmental management is not incompatible with environmental improvement. A visible reduction in pollution loads has been achieved in some Chinese cities. However, the approach has not, to any significant degree, strengthened local environmental capacity. Little attention has been given to strengthening technological, institutional or social capacities. Where financial capacity has been developed, it has been a consequence of the local ownership of financial resources.

The urgent need for broad based capacity building is further reinforced by the difficulties that donors face in implementing environmental projects within a transitional context. The major problem with Japanese environmental aid is that it targets the most polluting enterprises in China without taking into account the issue of industrial restructuring. How can Chinese enterprises reduce pollution and restructure at the same time without the necessary technological and managerial support? Under current conditions, the Japanese engineering approach to environmental management is more effective in China's poorer regions where industrial restructuring is less advanced. The question is, for how long is Japanese environmental assistance likely to remain effective when even poorer regions in China are under intensifying political and market pressures to reform SOEs?

In the face of a strong political mandate in Beijing, the ongoing reform of the banking system, the gradual establishment of private property rights, and China's recent accession into the WTO, scholars are increasingly optimistic about the future prospects for industrial restructuring in China.[128] It is, however, too early to determine the speed of economic transformation, particularly in China's poorer regions. In the meantime, environmental donors face a difficult decision over whether to continue to fund Chinese industrial enterprises. The fate of many SOEs in China remains unclear. Yet, these enterprises produce approximately 65 per cent of China's total pollution and, therefore, cannot be ignored. Local governments are also under pressure from the enterprises themselves to secure the necessary funding to enable them to meet national emissions standards and thus, remain in operation. Simply closing all of these enterprises down, as in the case of the Shenyang Smelter, is not a viable option. It would plunge millions of workers into destitution thus creating large-scale social instability.

In the future, the difficulty for donors will be in trying to distinguish between the potential winners and losers. A lack of transparency makes the task of evaluating the economic feasibility of industrial pollution projects particularly difficult. Nevertheless, to avoid investing unwittingly in non-performing enterprises both economic and social considerations will have to be taken into account. From the Japanese standpoint, future success in building local environmental capacity will greatly depend upon reconciling engineering efficiency with economic and social viability.

4 Managing the environment with a human face

The UNDP approach

In sharp contrast to Japan, the UNDP provides grant assistance aimed at developing human capacities rather than loans for building physical infrastructure. Projects are small, with limited amounts of funding. Capacity building is the primary goal of environmental assistance, as opposed to a mere by-product. The UNDP's approach to solving environmental problems is based on the belief that weak environmental management in developing countries is not simply caused by a lack of capital or technologies but by weak managerial capabilities. Human development that involves stronger participatory practices and technological innovation is, therefore, seen as essential.

The literature on the UNDP's environmental assistance is sparse.[1] Unlike Japan and the World Bank, the UNDP has not been a target of widespread external criticism, ostensibly because its environmental projects are small and less visible. UNDP projects are also specifically designed to meet domestic interests and needs and, therefore, are less prone to misplaced donor intervention. The fact that the UNDP has received far less attention (negative or otherwise) than other environmental donors does not necessarily suggest that its environmental aid is any less important. If carefully targeted, small-scale investments can lead to large-scale improvements. However, the size of the UNDP's aid projects does raise some interesting questions with respect to environmental capacity building. Can small investments in human development endure over time? Is a human development approach to environmental management sufficient in itself to strengthen local capacity? Or does it also require large-scale capital funding in order to ensure political commitment?

The findings in this case study suggest that, under certain conditions, the UNDP's environmental aid to China has enhanced local capacity. In some Chinese cities it has led to a significant improvement in institutional capacity by enhancing interagency coordination. It has also succeeded in demonstrating the importance of technological innovation, cleaner production, and to a lesser extent participatory practices. We shall see that the idea of participation can be adapted to Chinese circumstances, albeit in a more restricted way than first envisaged by the donor.

The UNDP's achievements in China are as much a reflection of the local commitment to protecting the environment, as they are an endorsement of the

UNDP approach. Moreover, the human development approach is not without its problems. The main constraint appears to be that, in the absence of enhanced financial capacity, any perceived or real improvements in environmental management are unlikely to endure over time. This cautions against the UNDP's current strategy of shifting its attention towards policy dialogue before address-ing the failings of its small interventions at the local level. Arguably, the power to influence Chinese policy lies in demonstrating the effectiveness of new ideas in practice.

I will begin this chapter by providing an overview of the UNDP's environmental aid programme as well as the motives behind its approach. Clearly, a human-centred approach to environmental management runs counter to China's traditional thinking and practice which raises the question of its acceptability in the eyes of the Chinese government. I will address this question by tracing the implementation process from policy dialogue at the national level to the local implementation of environmental projects in Shenyang, Wuhan, Benxi and Guiyang. To the extent that the human development model of environmental management can be applied to China, the key objective is to assess its effectiveness in actually building local capacity.

Towards global environmental advocacy

The UNDP, founded in 1965, is the United Nation's largest source of grants for development assistance. As a multilateral organization, the UNDP is linked to the United Nations General Assembly through the United Nations Economic and Social Council. The UNDP's development mandate has evolved over the past three decades from an original emphasis on basic human needs to the current paradigm of *sustainable human development*.[2] This has required considerable institutional adjustment that has not, as yet, been fully realized.

In the 1970s, the UNDP focused upon satisfying human needs by building self-reliance through technical assistance.[3] Environmental degradation was seen as a consequence of under-development which, in turn, focused attention upon the need for improving basic human needs. A decade later, growing international recognition of the importance of sustainable development, and the realization that environmental degradation was as much a result of over-development as under-development, meant that greater emphasis was placed upon supporting human needs without transgressing the carrying capacity of natural systems.[4] By the early 1990s, in the aftermath of the havoc wreaked upon developing countries by structural adjustment programmes, the UNDP distinguished itself from other donors by focusing upon development from a human perspective.

This new direction was heavily influenced by a book co-edited in 1989 by Richard Jolly (then deputy executive director of UNICEF), Giovanni Cornia and Frances Steward.[5] The UNDP's human development reports initiated in 1990 provided an important means of developing the necessary analytical tools for operationalizing the concept. Most important was the introduction of the human development index that symbolized a sharp departure away from conventional measures of wealth based upon GNP or per capita income by focusing upon indicators such as food security, educational performance and the ability to buy and sell basic goods and services.

At the same time, the UNDP became the main agency for coordinating efforts in sustainable development both within and outside of the United Nations system. Following negotiations at the UNCED in 1992, the UNDP took on prime responsibility for helping developing countries to integrate global environmental goals as outlined in Agenda 21 into their development plans.[6] To this end, the Capacity 21 fund was set up under the administration of the UNDP to principally fund national planning activities that involved stakeholder participation and broad-based information management.

In 1994, in order to integrate its human focus within the broader vision of the United Nations global agenda for sustainable development, the UNDP adopted a new development mandate that centred on *sustainable human development*. Under the administration of James Gustav Speth (former founder and president of the World Resources Institute), this was defined broadly as development that 'not only generates economic growth but distributes its benefits equitably; that regenerates the environment rather than destroying it; and that empowers people rather than marginalizing them'.[7]

Over the past decade, the continuing challenge for the UNDP has been to adjust its organizational mission to accord with its mandate. This has proven to be a difficult task for three predominant reasons. First, UNDP staff have traditionally been trained in project administration; they therefore have limited experience and skills in translating broad and arguably vague development goals into operational practice.[8] Second, the UNDP has been under constant pressure to support every development objective that is deemed important by the United Nations system as a whole – sustainable development, poverty eradication, the advancement of women, good governance and food security to name but a few.[9] Third, and perhaps most important, the UNDP has an annual budget of approximately US$2 billion, which is spread thinly over 174 developing countries. As the number and scope of development goals have expanded, the United Nations allocations for development funding have declined. Funding hit a record low in the year 2000 but the downward trend was later reversed in 2003 when total resources reached a high of US$3.2 billion.[10]

From the late 1990s onwards, the UNDP has carried out a series of administrative reforms. Two key structural changes have involved the establishment of comprehensive country programmes, and the delegation of programme approval authority to the UNDP resident representative. It was recognized that hundreds of small projects scattered around the world were at high risk of falling prey to local vested interests without a policy framework that was specifically linked to recipient concerns. With the arrival of a new Administrator, Mark Malloch Brown (previous vice-president at the World Bank), in 2000, further decentralization took place leading to a staff reduction of 25 per cent at the end of 2001. More recent reforms have aimed at improving development effectiveness by strengthening organizational capability in the three core areas of knowledge, partnerships for capacity development and advocacy. New innovations include the establishment of knowledge networks at the country level, the adoption of a new learning strategy for staff and the creation of an Ombudsperson Office to ensure greater accountability.[11]

These ongoing internal reforms have meant that the UNDP can now concentrate its efforts upon enhancing policy dialogue with developing countries rather than simply administering projects from a distance, or in UNDP parlance it can move towards 'upstreaming' technical assistance.[12] The UNDP's political neutrality relative to other donors means that it is in a good position to facilitate the national implementation of global environmental goals through policy dialogue. Given the political sensitivity surrounding the issue of environmental protection in developing countries, the UNDP's comparative advantage lies in its ability to develop policy dialogue based upon small interventions.

Addressing environmental problems is central to the UNDP's struggle to maintain a leading role in global development advocacy. Since the mid-1990s the UNDP has become increasingly marginalized within the donor community because it no longer enjoys a monopoly position in providing technical assistance; grant packages are now commonplace amongst a large number of donors (the World Bank included). Perhaps more important, as a consequence of the decline in its core funding allocations from developed countries, the UNDP increasingly relies upon global environmental funds such as the Global Environment Facility (GEF). This factor alone is a key determinant of the future direction of the UNDP's environmental assistance because it implies a shift away from local environmental concerns such as clean water and sanitation and a move towards global environmental concerns such as renewable energy and biodiversity conservation.

The scope of UNDP environmental assistance

Until the 1990s, the UNDP's environmental assistance was largely directed towards low-income countries in order to promote the management of natural resources. Then, under the broad rubric of sustainable development, attention shifted towards environmental governance (i.e. support for improved regulatory and legal frameworks, interagency coordination, stakeholder participation in decision-making and decentralization), sustainable energy, and environmental information systems. By 1997 the UNDP's environmental programme accounted for 24 per cent of total resources. Four years later this stronger focus on the environment was eclipsed by concerns over democratic governance which accounted for 45 per cent of total expenditure in 2001 compared with 14 per cent for the environment.[13]

Currently the UNDP's resources are allocated to five main thematic areas: poverty reduction, democratic governance, crisis prevention and recovery, energy and the environment and HIV/AIDS. This broadening of priorities can in part be explained by the need for convergence with the United Nations Millennium Development Goals (MDGs). At the Millennium Summit in September 2000 the international community made a commitment to work towards achieving significant improvements in people's lives and the elimination of poverty by 2015.

This included a focus upon equitable access to environmental resources and the broader goal of ensuring environmental sustainability.[14] In the words of Brown, 'the UNDP's vision is clear: to play a pivotal role as advocate, enabler and adviser, partner and leader in helping shift the MDGs from rhetoric to reality'.[15]

But changing priorities are also a consequence of the changing composition of the UNDP's resource base. Voluntary contributions to the UNDP's core resources (known in the UNDP jargon as Target Resources for Assignment from the Core (TRAC)) more than halved over the 1990s, dropping from US$927 million in 1995 to US$844 million in 1996 (which was largely the result of a 55 per cent cut in funding from the United States), and an estimated US$420 million in 2000.[16] Yet, during the same time period, contributions to non-core funding (trust funds, cost-sharing and government cash counterpart funds) increased bringing the total UNDP budget to approximately US$2 billion in 1996 and US$2.2 billion in 2000, including US$1.1 billion from OECD/Development Assistance Committee (DAC) donors and over US$900 million from cost-sharing government funds.[17]

Given the changes in the UNDP's financial resource base, the environment is likely to remain a top priority because of its central importance to non-core sources of funding. Non-core funds for environmental purposes include the Capacity 21 Fund, the Multilateral Fund of the Montreal Protocol and the GEF. Since its inception at UNCED in 1992, the Capacity 21 Fund has invested over US$85 million to assist developing countries with integrating the principles of Agenda 21 into national planning and development.[18] China was one of the first countries to receive funds for formulating its own Agenda 21 in 1993. The UNDP is the sole implementing agency for the fund. Although the fund is a crucial element of the UNDP's environmental programme, enthusiasm from member countries in providing funds for Capacity 21 has recently waned. Only Japan remains a committed supporter of the fund in providing a cumulative total of US$30.9 million in 2000, compared to a paltry US$1.9 million from the United States.[19]

The Multilateral Fund and the GEF have larger budgets. The parties to the Montreal Protocol established the Multilateral Fund in 1987 to assist developing countries in the control of ozone depleting substances (ODS).[20] The UNDP is one of the four implementing agencies together with the UNEP, the United Nations Industrial Development Organization (UNIDO) and the World Bank. The US$2 billion GEF fund also provides grants and concessional lending to assist developing countries in the reduction of ODS, in addition to reducing global warming, preserving biological diversity and protecting international waters. The UNDP is one of the three implementing agencies (with UNEP and the World Bank) and is also responsible for managing a small grant programme that supports community-based NGO projects.

The GEF, in particular, has had an important influence upon the UNDP's environmental agenda. According to UNDP officials in New York, by leveraging GEF funds onto core resources the UNDP is now 'playing a catalytic role' in energy-related environmental assistance.[21] In accordance with the energy-related principles of Agenda 21 (chapter 9), these programmes are geared towards the twin objectives of improving the efficiency of energy systems and developing renewable sources of energy. In general, energy is seen as a vital component of sustainable development and an important means of addressing all of the key objectives of the UNDP's development agenda.[22]

Bringing in the human dimension

The size and scope of the UNDP's financial resources are clearly important in shaping its environmental assistance. To attribute the UNDP's approach simply to financial pressures, however, would be misleading. Above all, it is the idea of human development that defines the UNDP's approach towards assisting developing countries with their environmental problems. Fundamental to the human-centred model of environmental management is a belief in the importance of managerial capabilities. The latter are conceived of broadly and include both institutional strengthening (i.e. government planning and decision-making) and technological advance that involves improvements in industrial management processes and practices.

With respect to institutional strengthening, participation is deemed to be essential. The UNDP draws upon the stakeholder concept of participation, which blends both the democratic and community perspectives on participation, as outlined in Chapter 1.[23] Stakeholder participation is seen as a functional means of ensuring that all relevant actors actively participate in policy dialogue and project implementation with minimal outside intervention. In keeping with its perceived role as a facilitator rather than a controller of development assistance, the UNDP has a strong preference for working within existing structures and institutions rather than imposing change from outside. Nevertheless, it cannot escape from the inherent contradiction in trying to develop participatory practices from outside rather than within. For the UNDP, stakeholders include national and local governments, civil society organizations, community-based organizations and aid beneficiaries.[24]

The UNDP's promotion of participation is by no means uniform. In the words of one UNDP official involved in developing participatory programmes in New York, 'participation has not yet been assimilated into the UNDP's culture and many UNDP staff still avoid using participatory methods'.[25] However, the recent shift towards 'upstreaming' UNDP development assistance has reportedly stimulated a stronger institutional interest in participation because it is perceived to be an important means of strengthening policy dialogue. Moreover, participation is seen as an essential prerequisite for the effective implementation of Capacity 21-funded projects and programmes.

The technological element of the UNDP's human development approach to environmental management is more widely embraced within the organization itself and can be applied with greater vigour because, unlike the participatory element, it is not subject to particular political arrangements in recipient countries. Drawing upon the conservative version of 'ecological modernization' theory, the UNDP firmly subscribes to the view that in order to mitigate both the cost-recovery problem involved in investing in environmental protection and the adverse effect of stringent regulations upon industrial competitiveness, industry needs to take no-cost or low-cost initiatives in improving process management. The UNDP's so-called 'barrier removal approach' assumes that small changes in industrial processes can lead to large impacts. For example, particulate emissions can be greatly reduced simply by adjusting the temperature of coal burning. In such a way, technological innovation can lead to substantial environmental and economic savings without requiring any technical upgrading.

Common to both participatory and technological elements is the belief that building upon human capital is the most appropriate means of ensuring that environmental projects endure over time and are ultimately sustainable. The approach is clearly relevant to China where environmental needs are great but investment and managerial capabilities are limited. Given the Chinese predisposition towards engineering solutions, the important question to ask is whether the approach could actually work in China. Before addressing this question let us first look at the nature of the UNDP's environmental assistance to China as well as the motives behind it.

The greening of UNDP assistance to China

The UNDP first provided technical assistance to China in 1978 for a total amount of US$15 million.[26] From 1981 onwards the aid relationship became formalized along the lines of five-year country programmes that were aligned to China's Five-Year Plans. UNDP assistance to China can be defined in four stages. The first stage, during the 1980s, was totally recipient driven, which resulted in poor quality control and subsequent project failure. Contrary to conventional wisdom, recipient ownership did not lead to more worthwhile projects in keeping with local knowledge and capabilities. Instead, a multitude of central government agencies exploited the funds for their own ends; projects were scattered according to the whims of particular officials and assistance became merely an exercise in what Engberg-Pederson and Jørgensen refer to as 'technical-gap filling'.[27] In the words of one UNDP official, formerly resident in China:

> Up until the early 1990s, UNDP projects in China were recipient oriented – if China wanted a project they got it – as a result, the UNDP has many museum pieces across China including a ship tanker for transporting coal and hundreds of boilers.[28]

The major focus of both the UNDP's first country programme (1981–1985) and second country programme (1986–1990) was upon industrial product improvement and agriculture. The environment was of limited concern, with the exception of a few projects that targeted industrial energy recycling and conservation, wastewater treatment and land resource management.[29]

During the second stage of UNDP development assistance to China (1991–1995) the UNDP was able to regain some control over its projects and programmes, largely as a consequence of its increased political profile following the UNCED Conference at Rio in 1992. Environmental issues provided a means for the UNDP to initiate a stronger policy dialogue with the central government in Beijing. Consequently, funding for the third country programme (1991–1995) was channelled into projects for improving energy efficiency, sustainable natural resource development and capacity building for the formulation of China's Agenda 21. The environment was not the only impetus for establishing stronger policy dialogue. Both sides were conscious of the fact that increasing foreign direct investment during this period had practically rendered obsolete the former strategy of using the UNDP as one of the few windows available for the transfer of international technology and industrial expertise.[30] The time had come, therefore, to redefine the UNDP's role in China.

The third stage, coinciding with the UNDP's first country cooperative framework (1996–2000), represented a watershed in the history of UNDP development assistance to China when, for the first time, a more equal partnership between donor and recipient was established. Largely driven by institutional restructuring within the UNDP, major shifts occurred in relation to agenda setting: projects became thematically driven (poverty eradication, environmental protection, agriculture and economic reform) rather than sectorally based; and the overarching objective became capacity building rather than improving the lot of a small group of beneficiaries.[31]

In 1997, the UNDP's broad sustainable human development mandate was translated into a more manageable strategy and action plan for the environment and sustainable energy. Within this framework, the four main areas included environmental governance (formulation and implementation of Agenda 21), sustainable energy development (efficient coal utilization, energy conservation and the reduction of greenhouse gas emissions), pollution prevention and control (predominantly air pollution control and the reduction of ozone depleting substances), and natural resources management.

Energy remains a common interest between donor and recipient, and it has also been a major focus of UNDP directed global environmental funds (Multilateral Fund and GEF). By 1999, China had received a total of US$60 million from the GEF together with US$57.6 million from the Multilateral Fund – the highest of any of the 33 recipient countries.[32] During the period between 1991 and 1999 the percentage of total UNDP funding allocated to environmental projects almost tripled. Although the small size of UNDP projects (with many costing less than US$1 million) makes an accurate calculation difficult, at a rough estimate, by 1999 the UNDP had provided approximately US$140 million in environmental assistance to China representing 20 per cent of total development assistance (US$695 million).[33]

The fourth and current stage of UNDP assistance to China is linked to the second country cooperative framework (2001–2005), which has been designed to strengthen convergence between China's national development priorities and the United Nations MDGs. Key programme areas include support for democratic governance, comprehensive poverty reduction, HIV/AIDS, sustainable environment and energy development, and crisis prevention and recovery. In the area of the environment, greater emphasis has been placed upon the development of renewable energy sources, improved pricing structures and the empowerment of local stakeholders. This new focus effectively espouses a pluralistic approach to environmental management that is likely to generate more positive outcomes in the future. Unfortunately, this has not been linked to environmental capacity building which is still narrowly defined in relation to the capacity 'to negotiate and implement global environmental commitments'.[34]

UNDP motives

The UNDP's motives in providing increasing amounts of environmental assistance to China are largely pragmatic. First, as one of the largest emitters of greenhouse gas emissions (though not in per capita terms), China is an obvious target for UNDP funding. By assisting the central government in developing policies to protect its environment, the UNDP is enhancing its profile as a significant political actor in global environmental advocacy. Moreover, by stressing the need for improvement in management capabilities, hitherto ignored by other donors, the UNDP is carving out a unique role for itself in the delivery of environmental assistance. In short, China provides a status-enhancing opportunity for the UNDP, made possible by the recent shift towards policy dialogue.

Second, as noted earlier, by providing environmental assistance to China the UNDP stands to increase its fiscal capacity. In essence, a strong presence in China represents a future hedge against further reductions in core financial resources. The UNDP can channel global environmental funds to China – a major recipient of both the GEF and Multilateral Fund – which can also be leveraged onto existing UNDP core programmes. Furthermore, China is a top recipient of international environmental assistance worldwide, providing the UNDP with an additional opportunity to put into practice its oft-cited objective of acting as a key facilitator and coordinator of international development funds.

A third motive, and no less pragmatic, is related to China's pre-existing capacity in environmental management which, albeit flawed, is still a major improvement upon most other developing countries. China's well-established regulatory framework and evolving environmental policies greatly reduce the risk of UNDP environmental projects becoming invisible – a particular problem of grant related development assistance in general. Indeed, in comparison to countries such as Somalia where absorptive capacity is exceptionally low, the hope is that environmental projects in China may actually make a difference.

Having examined the motives behind the UNDP's provision of environmental assistance to China, I can ask the next important question: how has the Chinese government responded to the UNDP's human development approach to environmental management? One might expect the UNDP to have some persuasive power with the Chinese government because it is seen as a 'trusted' international partner and, therefore, new ideas are more likely to be taken seriously. But on the other hand, as in the case of other international donors, the UNDP's political leverage in China is limited; no idea, no matter how important, is possible without the full endorsement of the Chinese government.

Environmental dialogue at the national level

Policy dialogue between the United Nations and China takes place directly through the UNDP representative office in Beijing with minimal interference from New York. Consequently, donor objectives are easier to reconcile with local needs and expectations. Environmental project negotiations have reportedly been relatively smooth.[35] This is, perhaps, not surprising given that technical assistance does not impose a heavy financial burden upon the recipient country; the total funding for each project is minimal – on average US$1 million with cost-sharing requirements of approximately 40 per cent.

The UNDP's designated counterpart agency in China is the China International Center for Economic and Technical Exchanges (CICETE) which is located directly under the Ministry of Foreign Trade and Economic Cooperation (MOFTEC).[36] The CICETE manages UNDP grants, promotes training exchanges with international development agencies and coordinates cooperation with international NGOs. In total, seven different agencies are involved in UNDP environmental project negotiations: CICETE, MOFTEC, NDRC, MST, MoFA, SEPA and SETC. The UNDP also holds parallel negotiations with bilateral donors to explore possible areas of coordination.

The administration of the UNDP's environmental assistance to China is made more complex by the involvement of other United Nations organizations (see Figure 4.1). For example, UNEP and the United Nations Centre for Human Settlements (UNCHS or Habitat) provide technical assistance for developing capacity in environmental planning and management at the municipal level in China through the Sustainable Cities Programme.[37] In addition, UNIDO provides technical support for industry specific environmental projects, especially those funded by the Multilateral Fund. The main problem with this rather cumbersome administration, however, is not so much on the donor side as on the recipient side with agencies such as the NDRC, SETC and SEPA all jostling for position over administrative turf and budgetary allocations.[38]

As discussed earlier, in the first country cooperative framework in 1996, overall policy objectives had to conform to China's Five-Year Plans and, at the same time, meet the requirements of the UNDP's sustainable human development mandate. To date, the priorities of donor and recipient have differed over means rather than ends. For example, the sustainable energy objective represented a perfect match between the UNDP's new focus of expertise upon sustainable and renewable energies and China's insatiable demand for energy as the engine of future economic growth. But at the onset of negotiations, in order to achieve this goal the Chinese stressed the importance of science and technology whereas UNDP officials were keen to promote enhanced participation in national development activities – including the participation of women.

Apparently a consensus was eventually reached fairly easily, which is more remarkable when one considers that the UNDP was embarking on uncharted terrain. Up until that point, the issue of participation had been off-the-aid-policy agenda; it was considered by donors to be contrary to China's political orientation

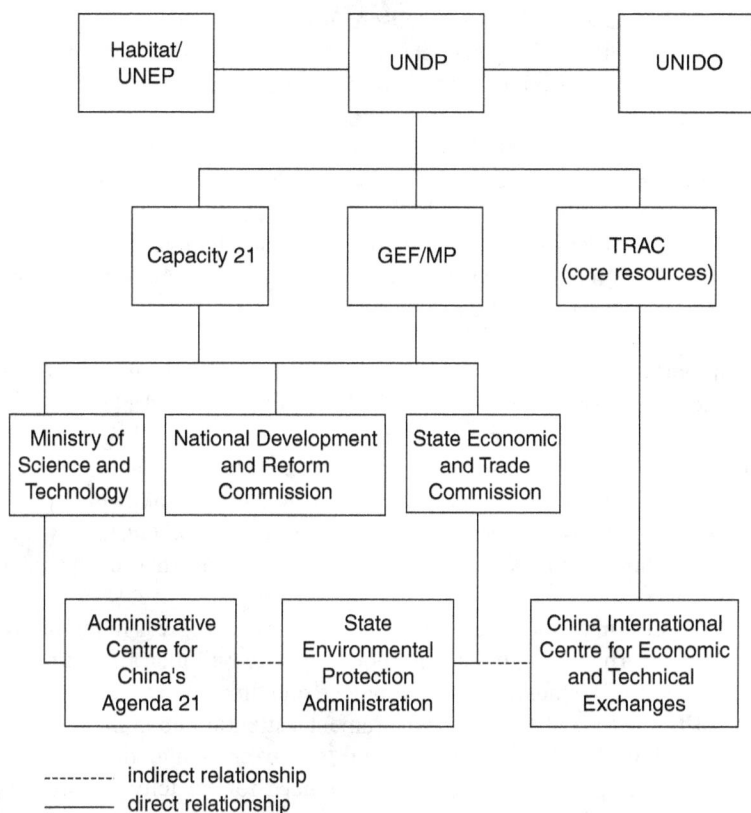

Figure 4.1 The administration of UNDP environmental aid to China.

Source: Based on interviews with the UNDP representative office in Beijing.

and, therefore, unlikely to be taken seriously by the Chinese government. But the real barrier to adopting a participatory approach, at least in theory if not in practice, was less political or conceptual and more practical. In the words of one UNDP official involved in the negotiations at the time:

> In fact, China was not difficult to persuade over participatory style arrange-ments because they are used to working that way with leading groups and working councils – decision-making in China is slow and iterative. The problem was more one of resource allocation than concept. When many parties and stakeholders are involved as opposed to one single counterpart agency, the sharing of budget allocations is difficult to negotiate.[39]

China's acceptance of the notion of stakeholder participation can best be illustrated by the fact that it was the first developing country to formulate its National

Agenda 21 in 1994, involving the coordination and participation of 52 different agencies, departments, and mass-based organizations in Beijing, together with 300 national experts.[40] The final chapter of China's National Agenda 21 is devoted to the issue of public participation in sustainable development and is framed as follows:

> New mechanisms are needed for public participation in sustainable development. It is necessary for the public not only to participate in policy making related to environment and development, particularly in areas that may bear direct impact on their living and working communities, but also to supervise the implementation of policies.[41]

It is important to stress, however, that in actuality the Chinese version of participation at the national level was undeniably elitist. Limited attention (if any) was given to the importance of consulting the Chinese public at large. This was a direction which the UNDP at the time was unwilling to pursue. Instead, it preferred to remain politically neutral, epitomized in the following clause in the UNDP Cooperative Framework: 'the UNDP assists, *at the request of* the Chinese government, in building capacity for good governance and popular participation'.[42]

Despite these constraints, the UNDP has made some progress in familiarizing the Chinese government with the concept of capacity building. In particular, its assistance for the formulation, development and implementation of China's Agenda 21 has helped to focus the government's attention upon the importance of strengthening stakeholder participation.[43] The challenge that the UNDP now faces is to enhance understanding of the need for capacity building at the local level.[44] This requires greater efforts to strengthen the effectiveness of its environmental projects. But this, of course, runs counter to the UNDP's more recent emphasis upon strengthening policy dialogue at the national level. In this respect, an investigation into the local implementation of environmental projects in China has important implications not only for the UNDP's environmental programme but also for its evolving policy towards development assistance per se.

The remainder of this chapter will focus upon the local implementation of UNDP-funded environmental projects. Special attention will be given to addressing three key questions. How is the UNDP's approach to environmental management perceived at the local level in China? Under what conditions is it effective in enhancing participatory practices and technological innovation? And, to what extent is UNDP environmental assistance building local capacity?

Implementing environmental projects at the local level

One of the main constraints upon local capacity building in China is the lack of enthusiasm at the national level. According to one UNDP official, the central government is not interested in developing capacity in the hinterland because 'the control strings are longer and the resources are less'.[45] Indeed, up until the late 1990s, the majority of UNDP capacity building projects and programmes were highly centralized. In 1996, out of the 15 environment and energy projects initiated through UNDP core funds more than half were located in Beijing, with only one project located in the poorer region of Guizhou.[46] The projects investigated in this case study, therefore, not only set a precedent in relation to participation, but are also amongst the first UNDP capacity building projects in China to be located outside of Beijing.[47]

The selected projects reflect both the participatory and technological elements of the UNDP's approach to environmental management. They include 2 sustainable city projects in Shenyang and Wuhan, and 2 air pollution projects in Benxi and Guiyang. All 4 projects draw upon the UNDP's core resources rather than global environmental funds so that the UNDP remains central to institutional involvement. I should note, however, that the sustainable city projects are part of a broader global programme for sustainable cities under the auspices of UNEP and Habitat.

Three of the project sites, Shenyang, Benxi and Guiyang, overlap with the Japanese project sites discussed in Chapter 3 and, therefore, also provide opportunities for comparative analysis between the relatively developed region of Liaoning province and the poorer region of Guizhou province. Wuhan, located in Hubei province, is another example of a Chinese municipality that has a relatively high level of economic development. Situated in the middle reaches of the Yangzte River, Wuhan has made considerable economic progress over the past two decades by acting as a bridge between China's booming coastal cities and nine inland provinces.

UNDP sustainable city projects

The UNDP sustainable city projects in Shenyang and Wuhan began implementation in 1997 and were completed in 2000. Initiated by the UNEP/Habitat Sustainable Cities Programme (SCP) in 1994,[48] the UNDP took over control in 1996 when UNEP ran out of funds. Investment was modest, amounting to a total sum of US$714,500 for each project (the UNDP provided US$400,000, the municipal government a further US$260,000, and UNEP only US$54,500).

The SCP aims 'to strengthen local capacities for strategic planning and environmental management based on the active participation of the municipality and its partners in the public, private and community sectors'.[49] Thus, the emphasis is upon a paradigm shift in urban environmental management away from traditional technocratic master planning and towards broad-based stakeholder participation. The underlying assumption is that in solving environmental problems cities must first address environmental planning and management constraints. In short, 'cities face a scarcity, not of capital, but of management capacity'. Ultimately, given the fact that traditionally a large number of agencies are involved in urban environmental management at the municipal level, a major concern is to promote 'interest reconciliation' and to institutionalize the SCP process by integrating it into mainstream decision-making.[50] An overlap exists with Capacity 21-funded programmes in that SCP demonstration projects, in a formal sense, are expected to lead to the formulation of a local Agenda 21.

The SCP process is composed of four interrelated phases. The first phase involves the preparation of an environmental profile as the foundation for an environmental information system. The second phase concentrates upon identifying key environmental priorities through a city consultation between private, public and community stakeholders. Phase three involves the establishment of working groups, comprised of representatives from key government agencies, to negotiate strategies on the basis of cross-sectoral participation, and phase four focuses upon institutionalizing the management system. Throughout the process technical training workshops are held in environmental risk and environmental technology assessment methodologies.

The SCP is especially relevant to China where typically nine agencies are involved in urban environmental services at the municipal level. Interagency coordination is a particular problem because of the continuing legacy of central planning and sectorally based investment projects. In general, the main responsibility for interagency coordination falls upon the Municipal Planning Commission, with leading groups made up of key relevant agencies to improve coordination on specific issues.[51] Despite these efforts, according to the UNDP: 'each institution still generally proceeds independently, collects its own information, and creates its own programmes following the general plans and principles established by the central leadership.'[52] In 1996, the UNDP project assessment report identified five major limitations with urban environmental management in China: poor interagency coordination, a lack of public participation, weak strategic development planning, inefficient financial management, and limited individual skills and experience.[53]

The SCP projects in Shenyang and Wuhan aimed to address all of the earlier capacity deficits. More specifically, the projects had four key objectives:

1   To establish cross-sectoral working groups as a continuing mechanism for municipal government decision-making.
2   To integrate environmental strategies and planning into a citywide environmental strategy based upon Agenda 21 principles.
3   To formulate 'bankable' capital investment projects to attract external funding.
4   To provide an important model of environmental management for replication in other Chinese cities with the support of the ACCA21 in Beijing.[54]

CICETE maintained overall authority over the projects and a small office within the municipal EPB at each project site was responsible for actual project implementation. As in the case of Japanese environmental projects, a leading group composed of high-ranking officials from key municipal government agencies (including the Municipal EPB, Municipal Planning Commission, Municipal Science and Technology Commission and the Municipal External Trade and Cooperation Commission) was set up to oversee the implementation process.[55] Throughout the project cycle technical support was provided by Habitat and UNEP's International Environmental Technology Centre.

The projects in Shenyang and Wuhan pursued the same SCP process and followed the same training programmes; hence both projects had similar objectives,

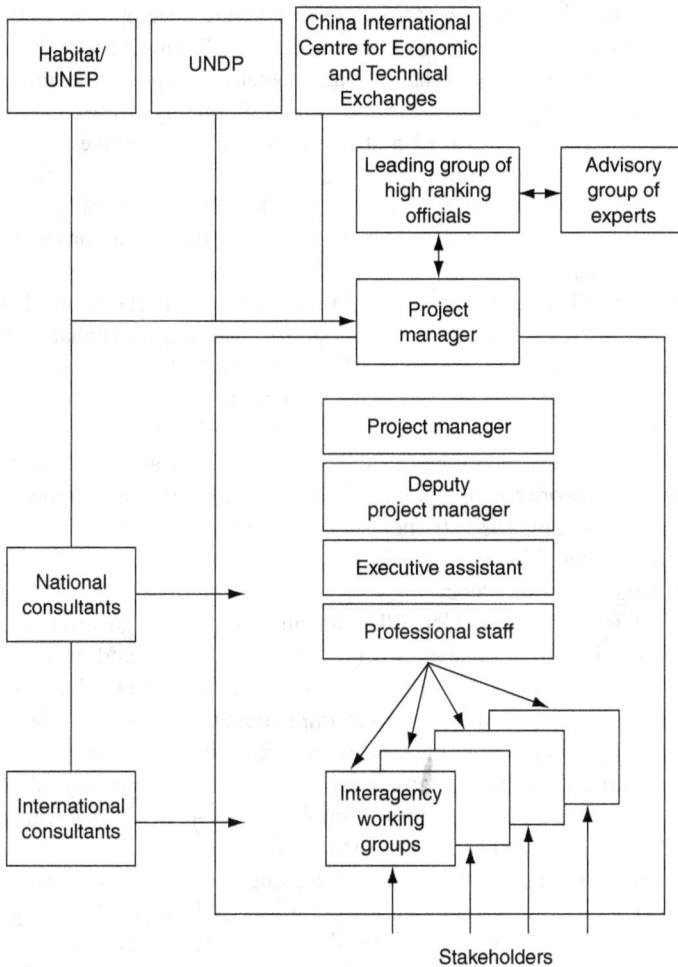

Figure 4.2 The structure of the UNDP sustainable city projects in China.

Source: Adapted from UNDP China, 'Managing sustainable development in Shenyang', Project No. CPR/96/321/A/01/99, unpublished, Beijing: UNDP, 1996.

structures and procedures (see Figure 4.2 for an overview of the project structure). However, the political and institutional arrangements varied, especially in relation to the political status of those officials participating in the project which, in turn, regulated the degree of local political commitment.

Managing sustainable development in Shenyang

Shenyang was the first city in China to join the United Nations Sustainable Cities Programme in 1994. According to EPB officials in Shenyang, the city was selected because it symbolized on a grand scale the challenging task which cities all over China faced in attempting to pursue sustainable development while still being overly dependent on heavy industry. Moreover, Shenyang was deemed a suitable location because of its relatively high level of economic development, strong environmental regulatory framework, and high quality research institutes. The selection criteria also involved a strong commercial incentive. The SETC believed that the SCP would ultimately enhance Shenyang's potential for developing environmental industries and services. In the 1990s, Shenyang had been selected by the central government as a lead city in China for the production of wastewater equipment.

But above all, Shenyang was chosen to be China's first United Nations sustainable city because SEPA in Beijing together with the provincial governor and the mayor of Shenyang were reportedly confident that 'Shenyang had both the political commitment and the necessary capacity to implement the project'.[56] Shenyang was one of China's first cities to establish an environmental protection bureau in 1981 and since then it has established a relatively advanced regulatory and legislative framework for environmental protection. Environmental monitoring capacity is strong in comparison to most cities in China.[57] In addition, the quality of the EPB staff (total 130) is relatively high both in relation to technical skills and English language competency.[58]

Project negotiations between UNEP/Habitat and the Chinese government at the central, provincial and municipal levels lasted two years and were stalled temporarily due to budgetary constraints on the donor side. The UNDP agreed to fund the project in early 1996 and formal implementation began in May 1997. From the very beginning, the municipal government displayed a high degree of political commitment. A leading group for the project was established on the basis of the former leading group for the implementation of China's Agenda 21 formed in 1994. The incumbent Mayor, Mu Suixin, a keen advocate of sustainable development, was appointed chair of the group. A project office was set up with an initial staff of 6 (which later grew to 9) seconded from the Shenyang EPB. Of critical importance was the fact that the deputy secretary-general of Shenyang municipal government was appointed as project director. This meant that because of his standing, the Municipal Planning Committee (which traditionally in China has greater political leverage than the Municipal EPB) could not overrule project implementation.[59]

In accordance with the SCP process described earlier, the first phase of the project involved the preparation of an environmental profile of Shenyang to provide an information base for future environmental priorities and strategies. To this end, UNEP and Habitat provided the technical support to establish a simple Geographic Environmental Monitoring System. Consultations took place with relevant government agencies, enterprises and experts. The environmental profile, published in April 1998 and widely distributed amongst the general public, identified three key priority issues: a shortage of water resources and serious water pollution, serious air pollution caused by the burning of coal, and serious domestic waste pollution. The profile also noted that public organization in

Shenyang with respect to environmental issues was limited. Key NGOs included the Municipal Women's Federation, the Municipal Institute for Environmental Sciences, the Federation of Trade Unions, and the Association of Science and Technology. It was acknowledged, however, that these organizations 'still have a considerable governmental flavour'.[60]

At the City Consultation, which took place during 5–7 May 1998, papers written by local experts were presented on three key environmental priorities. More than 300 representatives from government, enterprises, research institutes and so-called NGOs attended. The latter included the industrial department of the Shenyang daily newspaper and the people's government of Xinchengzi district. But despite the lack of genuine public participation, political participation was high – 86 representatives participated in the three day conference (mostly at the vice-director level) from all of Shenyang's leading government agencies together with representatives from the people's congress and the political consultative conference.[61]

During the second phase of the project, five working groups were established in the following areas: water resources, air quality, solid waste, cleaner production and city greening. The working groups consisted mainly of government officials from the relevant departments, together with some academics. The purpose of the working groups was to bring about innovation within existing institutions rather than to create a new institution for environmental management; they acted more like advisory councils for respective government departments by providing a mediating function for the reconciliation of vested interests. The main concern was to disaggregate overall strategies into viable and workable plans for implementing sustainable environmental projects.

By May 1999, to demonstrate their expertise in cross-sectoral management, the working groups had implemented a total of seven demonstration projects (with local funding) including the construction of a garbage station, the introduction of cleaner production processes in heavily polluting industries, and the construction of 100 public toilets. The latter project was seen as a particular victory: up until that point divisions between the Bureau of Construction and the Bureau of Public Health had meant that public sanitation had not kept apace with the city's rapid urbanization, with some residents having to cycle for five minutes simply to use a communal toilet.[62]

In the third and final phase of the project cycle, these demonstration projects were presented at an international environmental strategy workshop convened in Shenyang between 5 and 8 October 1999 with the explicit intention of displaying local capacity in environmental management in order to attract external investment for priority projects. A second conference was held a few months later for a domestic audience to consider ways of replicating the demonstration projects in other parts of China.

Overall, the success of the project was most marked in relation to improved stakeholder participation at the elite level. Its apparent failure, in contrast to other sustainable city projects worldwide, was in the lack of enhanced public participation. That said, in the final analysis, local rather than outsider perceptions matter most.

During interviews, project officials concluded that both governmental and public participation had increased. They maintained that interagency coordination had improved greatly but that the process had not been without its problems. In the words of the deputy project manager:

> In one sense it was relatively easy to mobilize participation amongst different bureaucracies as a strong will existed to take part in consultations. The project has been very effective in regard to training programmes, workshops, and enhanced awareness of stakeholder participation amongst government departments. On the other hand, it has been more difficult to resolve conflicts between sectoral interests relating to environment and development. What has been most important is to arrive at a clear understanding that the EPB is not the only agency responsible for environmental protection but, instead, plays the role of a facilitator in consensus-building.[63]

Equally important was the fact that participants in the project now viewed environmental management as 'process-oriented' rather than as a multitude of 'end-of-pipe' solutions. Moreover, the importance of managerial capabilities or capacity building (*nengli jianshe*) was becoming ingrained in the thinking of government stakeholders. This changing mindset was not confined to officials in Shenyang. The provincial government had also become aware of the need to focus more upon management rather than simply technology and funds. The governor of Liaoning province, Wen Shizhen, stressed the need to enhance managerial capabilities in a speech at the Provincial Environmental Protection Work Conference held in Shenyang from 1 to 2 August 1997 as follows:

> It should be noted that some problems regarding pollution control result from insufficient funding investment, while others are from poor management. This, we must overcome – the tendency of 'laying emphasis upon control but not on management'. Truly intensifying management and supervision over the environment can also yield good results.[64]

Aside from positive changes at the elite level, officials also maintained that the project had helped to stimulate public participation. In particular, they claimed that public awareness had increased as a result of the efforts made by the project office to disseminate information. In addition, platform conferences such as the SCP annual conference[65] (which took place in Shenyang from 30 September to 4 October 1997) and the City Consultation had helped to stimulate public interest. On both occasions the media were present, schools participated, and conference proceedings were broadcast through Shenyang's local television station. As a consequence, according to one EPB official: 'local people now have a clear idea of what sustainable development means.' But in effect, the concept of participation has proven malleable on the ground. In the eyes of local officials top-down participation is still seen as the most effective means of solving environmental problems based on the assumption that when the political leadership is committed the people will follow.

One of the perceived benefits of the project overall has been the increase in political commitment within the higher echelons of the municipal government. The mayor, in particular, has not only confirmed his commitment towards solving the city's environmental problems in words but also in practice – he has reportedly made personal visits to all of the 100 factories in Shenyang that were in the process of implementing cleaner production. In response to his visible political commitment to the project, the mayor was awarded an honorary prize from Habitat in 1998. The words used to describe the event in an EPB report in May 1999 strongly convey the sense that high level political engagement was seen as the key to public participation:

Mr Mu Suixin, the mayor of Shenyang, was awarded the Habitat Scroll of Honour in 1998 by the UNCHS because of his great contribution to the implementation of the Shenyang Sustainable Project. It was not only the honour of Mayor Mu, but also the honour of the Sustainable Shenyang Project as a whole, and it greatly strengthened the decision of 6,700,000 civilians to implement sustainable development in Shenyang.[66]

Although the issue of public participation was stressed in all of the project reports, in private, local officials were not convinced that inviting the public to participate in decision-making would enhance the quality of the decisions being made. In their view the process would become too cumbersome. Local officials preferred to subscribe to a form of utilitarianism in which the government thinks about the greatest benefit to the greatest number as a means of resolving conflicts of interest. Put succinctly in the words of one EPB official: 'in the event that the local government needs to widen a road and some resident housing is in the way then the residents will have to be relocated. The government cannot take the minority view.'[67]

This does not mean, however, that the idea of providing a voice for the people is always abandoned. Indeed, aside from traditional representation by delegates from the people's congress, local citizens can have their say on the issue of environmental protection through a 'complaints' hotline connected to the local EPB.[68] Moreover, the municipal government has been making increasing efforts to conduct public opinion surveys. For example, in 1998 a survey was conducted to assess the effectiveness of government activities in 16 key areas that involved 170,000 households. Environmental reform was given first place with a 94.5 per cent satisfaction rate.[69] Surveys of popular sentiment are arbitrary by nature. Nevertheless, the fact that such activities are being carried out reflects a sense of public accountability, if not public participation. In the words of the deputy project manager: 'The government system must be put under the supervision of the public' – an echo of the official credo in Beijing.[70]

From the local perspective, the only major reservation with the project was in relation to financial management. The general feeling amongst the project staff was that environmental plans had much improved but that they could not afford to implement them. Local resources were limited and in order for the project to be fully effective substantial funds were required from outside of China.[71] One positive signal in this respect was a commitment from the European Union to fund an integrated urban environmental programme in Shenyang at a total cost of US$400 million in grant assistance (a sum considerably larger than the US$4 million provided by the UNDP).[72]

The European project has six components (water resources, air pollution, cleaner production, energy, industrial restructuring and investment promotion) under the supervision of the provincial EPB, and one component (urban planning) under the supervision of the Shenyang EPB. It was significant that at the early stage of implementation those local officials responsible for implementing this project were not aware of the nature of the UNDP project and, indeed, the European Union representative in Beijing was equally in the dark. The local manager of the European Union project claimed:

> The main difference between the urban planning component of the European Union project and the UNDP project is that the UNDP project is only about capacity building, training and conferences etc. It is not about developing an integrated master plan by looking at all sectors and merging them into one.[73]

Such lack of communication and understanding between local officials ostensibly involved in the same activity highlights the fundamental need not only for funds per se but also for the improved coordination of funds. The implication here is that to avoid duplication of activities the SCP projects need to promote better coordination with other donors.

Project officials also expressed a minor reservation in relation to the United Nations' emphasis upon linking local capacity building with a global partnership between SCP cities worldwide. They claimed that it was necessary to learn from more advanced countries rather than other developing countries because cities such as Dar es Salaam in Tanzania did not even have in place the basic foundation of environmental management – a comprehensive regulatory framework and legislation. On the other hand, they were willing to concede that Shenyang's experience in implementing the project was highly relevant to other cities in China – a major affirmation of the project's worth overall.[74]

To summarize, in general, the UNDP's environmental project in Shenyang was relatively successful. Although the project did not succeed in stimulating public participation, it did facilitate an elitist form of stakeholder participation by improving interagency coordination. Local officials also became more aware of the need for capacity building in order to solve environmental problems. What is interesting from a comparative perspective, is that the UNDP's sustainable city project in Wuhan proved much more difficult to implement.

Managing sustainable development in Wuhan

Wuhan (population 7.2 million) is 1 of China's 6 largest cities. It was the first major inland city in China to open its economy to the outside world and, as a consequence, achieved consistent high rates of economic growth (average 15 per cent) throughout the 1990s. As in the case of Shenyang, industry is responsible for a large share of Wuhan's environmental problems, other important causal factors include heavy traffic, a relatively high population density (the third highest in China with 839 people/km^2)[75] and traditional eating practices.

Compared to Shenyang, the air quality in Wuhan is relatively good: SO_2 emissions are below the national average and TSP emissions are around 200 $\mu g/m^3$. But tail gas emissions from automobiles are causing a problem – the highest daily average concentration of NO_x emissions in 1996 was 356 $\mu g/m^3$, exceeding the national average by seven times.[76] Municipal solid waste (residential and commercial) is also a problem, although less severe compared with the cities in the northeast of China. With respect to air, water and solid waste pollution, the rapid growth of the catering trade poses a particular environmental hazard. Traditionally, residents in Wuhan prefer to eat breakfast outdoors, which has led to a proliferation of thousands of uncontrolled snack stalls using low quality coal and creating vast quantities of dirty water and waste.[77]

In contrast to the water scarcity problem in the northeast of China, Wuhan – situated at the intersection where the Yangzte River meets its main tributary the Hanshui River – suffers from serious flooding. Every year during the rainy season from July to August flood levels on the Yangzte River can reach as high as 27 metres.[78] Water pollution is no less serious. Wuhan has become known as the 'industrial black belt' of the Yangzte River (with 8.8 million tons of industrial wastewater and 3.3 million tons of domestic wastewater dumped into the river annually). The East Lake (Donghu) suffers from severe eutrophication. Moreover, a combination of factors such as illegal land reclamation practices, a high population density and the use of construction materials such as cement and asphalt that retain heat easily, mean that Wuhan is now experiencing its own 'local global warming' effect. During the summer months central Wuhan is now two degrees hotter than the outskirts of the city.[79]

The Wuhan municipal government has made some progress in dealing with its environmental problems but, like other cities in China, remains dependent upon external assistance.[80] UNDP environmental assistance to Wuhan can be traced back to initial discussions held in 1995 between UNEP/Habitat and the central and municipal governments over solid waste information management – at the time Wuhan had been selected by the central government as a priority city for improving solid waste.[81] This had more to do with commercial aspirations than political or institutional viability. A memorandum of understanding was signed and later withdrawn when the UNEP representative retired. Wuhan was then selected by the central government as China's second sustainable city under the UNDP. The SETC reportedly perceived the UNDP project to be an important bridge to attract private investment in environmental industry and services. Wuhan was also selected because of its relatively high economic development and good quality universities and research institutes (Wuhan ranks third in China, after Beijing and Shanghai, in scientific research and education).[82]

Yet, from its very beginnings in May 1997, the UNDP project in Wuhan lacked the strong sense of political purpose and institutional capacity that was apparent in its sister project in Shenyang. The municipal government placed the project under the supervision of Wuhan's existing leading group for implementing China's Agenda 21, chaired by the Vice-Mayor, Wu Houpu. A project office with six staff was set up in the EPB with the deputy-director of the Planning Committee appointed as project director.[83] His political standing was far lower than his counterpart in Shenyang. Most important was the fact that the city mayor was not actively involved in the project.

Unlike the situation in Shenyang, the EPB in Wuhan has a relatively low political status and remains under the direction of the Municipal Construction Bureau. Not surprisingly, the project office experienced considerable difficulty in galvanizing various governmental departments into action in order to implement the project. In the words of the deputy project director, 'Our office and EPB does not have the power to persuade different government agencies to participate.'[84] To overcome this barrier, rather than attempting to retain complete responsibility within the EPB, the project office delegated certain tasks and responsibilities to different agencies such as the Health and Sanitation Bureau. Nevertheless, interagency coordination was slow and the project office remained very much behind the scenes.

But at least in relation to the management of environmental information, the UNDP project was more successful in Wuhan than in Shenyang – a reflection of the high quality of Wuhan's environmental research institutes. From a technical perspective, the 'Environmental Profile' was effective in covering Wuhan's environmental problems in a comprehensive and detailed manner, although it failed to come to grips with some of the difficulties involved in environmental management: less emphasis was put upon interagency coordination and more upon legal frameworks and science and technology support. In addition, no mention was made of the need to enhance public participation. The City Consultation that followed involved a large number of government agencies, but community-based groups were few and far between. Those that attended included the Workers Union, the Women's Federation, the Industry and Trade Union and the Students Union.[85]

Like their counterparts in Shenyang, local officials were sceptical of the need to increase public participation beyond existing mechanisms in order to improve environmental management. The manager of the project office believed that actual public representation was inherently difficult. In his words:

> Sometimes it is not possible to wait to get a public consensus on issues such as phasing in cleaner coal or introducing cleaner transport. We have to establish the regulations and citizens must abide by them. Already we pay a great deal of attention to encouraging public participation in environmental management.[86]

In Wuhan, the EPB had its own independent environmental propaganda and education office; the Wuhan government publishes an environment paper, *Xinfan*, twice a week; and a 'complaints hotline' is in operation 24 hours a day. Most complaints focus upon the catering trade, construction noise and the pollution of the Donghu Lake where the city's major universities and research institutes are located. The Internet is also eliciting entirely new levels of public participation with many complaints and suggestions now being transmitted via email. This new form of communication is important; it represents a psychological break with the notion of 'propaganda' – or the one way heavily censored flow of information – and a move towards a more democratic and interactive means of communication. That said, access to the Internet is still limited and unlikely to make a major impact on environmental management in the short term.

At the City Consultation between 28 and 30 April 1998, participants agreed to focus upon four priority environmental issues: water pollution control, air pollution control, solid waste management and cleaner transportation. Working groups were then formed in these key areas. But in comparison to Shenyang, demonstration projects lagged far behind.[87] This was largely a result of three interrelated factors: first, interagency coordination was still weak; second, the Wuhan project office was collaborating closely with the Municipal Foreign Economic and Trade Committee (FETC), which was strongly oriented towards attracting investment as opposed to demonstrating capacity;[88] and third, Wuhan officials, like their Shenyang counterparts, were concerned that improvements in interagency coordination, no matter how great, would flounder if action plans could not be realized for lack of funds. Thus, the third phase of the UNDP project in Wuhan ended on a traditional note: the Sustainable Development Financing Conference, held in Wuhan between 23 and 25 May 2000, was primarily focused upon soliciting external funds for 27 major projects for environmental investment.[89] Limited attention was given to demonstration projects, or the city's capacity in environmental management. There was, seemingly, little difference between this conference and countless other international conferences held in China to attract private investment.

In comparing the implementation of the UNDP projects in Shenyang and Wuhan, the previous discussion suggests that a human development approach to environmental management is unlikely to work without pre-existing political commitment and institutional capacity. Moreover, in order to endure over time it would seem that efforts to build managerial capacities also require access to recurrent funding. The critical importance of funding was also highlighted in the evaluation report prepared by Habitat in Nairobi. It reported that 'the discussions held with the two SCP projects show that the requirements of funds for future sustainability will be immense'.[90] The report also noted the relative success of the SCP project in Shenyang compared with the project in Wuhan. Yet it gave little indication as to why this was the case. According to the Habitat advisor to the project, differing perceptions were a critical factor in that the Wuhan project office 'tended to treat the whole exercise as a project rather than a process'.[91]

Let us now turn to examining the implementation of the UNDP's air pollution projects in Benxi and Guiyang. Here the emphasis is upon technological innovation and strategic environmental planning rather than stakeholder participation. What is interesting from a comparative perspective is that these projects faced the same problem with respect to political commitment and institutional capacity.

UNDP Air Pollution Programme

The UNDP's Air Pollution Programme, initiated in 1997 and completed in the year 2000, aimed to promote a unified approach to air pollution management in China on the basis of technological innovation. The programme represented a microcosm of the major air pollution problems facing China: SO_2 from coal combustion in Guiyang, TSP from anthropogenic and natural sources in Xi'an, NO_x from automobile emissions in Guangzhou and generic emissions (including CO_2) from industries in Benxi.

The main objective behind all four pilot city projects under the programme was to assist local governments in the development and enforcement of environmental policy by identifying and monitoring sources of pollution. In Guangzhou, for example, it was not yet politically feasible to implement a strategy for dealing with traffic pollution until the origins of NO_x emissions could be determined *vis-à-vis* Hong Kong.[92] In addition to improving information and monitoring systems, the UNDP aimed to enhance the capacities of local governments in environmental decision-making. To this end, the programme included a national support project, located within CICETE, to consolidate the experiences of the four pilot cities into a unified framework for air pollution control in China.

CICETE was given overall responsibility for the programme, and various research institutes (including the Chinese Research Academy for Environmental Sciences (CRAES), the Chinese Academy of Science (CAS) and the Centre for Environmental Science at Peking University) provided the necessary technical support. The ACCA21 office in Beijing was responsible for the dissemination of environmental management practices throughout China. The two project studies that follow focus upon the cities of Benxi and Guiyang. Although the projects pursued similar objectives geared to technological innovation, implementation varied as a result of differing political and institutional arrangements.

Capacity building for widespread adoption of clean production for air pollution control in Benxi

The UNDP project in Benxi addressed the critical issue of how to pursue the transformation of a local economy with a disproportionate number of heavy industries in a cost-effective *and* environmentally sound manner. Benxi was one of China's 10 pilot cities selected by the central government for the implementation of its National Agenda 21. In June 1994, the municipal government published its own Agenda 21 with the two central objectives of developing cleaner production and reducing the city's dependency upon heavy industry.[93] The UNDP agreed to support Benxi at an international environmental meeting held in Beijing in 1996 when representatives from the municipal government presented a small number of cleaner production projects.[94]

At a total cost of just over US$1 million (with US$530,000 provided by the UNDP and the remainder from the municipal government), the UNDP project had three main objectives: to identify appropriate cleaner production technologies for Benxi; to propose alternatives for industrial restructuring; and to strengthen the environmental decision-making capabilities of the Benxi municipal government. In accordance with the project feasibility assessment carried out by the UNDP, the project also aimed to identify a financial mechanism for conducting environmental audits and to assess the feasibility of establishing a Cleaner Production Centre in Benxi.[95] The Benxi Agenda 21 Management Office was given the lead role in implementing the project (with 4 permanent staff and 5 temporary staff) under the dynamic leadership of Madam Tang Guimei – former deputy-director of the Municipal Science and Technology Commission. A leading group consisting of 14 government agencies was set up to oversee project implementation under the chairmanship of the city mayor.

By May 1999, at the time of my visit to Benxi, substantial progress had been made in the promotion of cleaner production. The project office had carried out an environmental audit involving 15 (out of a total of 500) enterprises in Benxi with support from international consultants, the National Center for Cleaner Production in Beijing and six research institutes in Liaoning and Beijing.[96] To avoid the risk of investing funds into enterprises that faced imminent closure, enterprises had been selected on the basis of profitability, together with a willingness to participate. The audit identified a range of no-cost, low-cost and high-cost cleaner production measures. At three of the factories involved, the economic and environmental benefits of cleaner production were already apparent (see Tables 4.1 and 4.2).

Despite differences in ownership (the Benxi Cement Factory is now fully privatized), all three factories had managed to reduce pollution loads and simultaneously recover some, or all, of the costs involved. According to the director of the project office, after only one year the economic benefits from all 15 enterprises had totalled RMB13.72 million (US$1.65 million).[97] Consequently, the response from some of the enterprises involved had been extremely positive. In the words of one enterprise manager:

> Management awareness has been gradual and at first driven by the government. But the results from the audit have greatly exceeded our expectations, so we would like to develop our own cleaner production audit and apply it to daily operations at the factory.[98]

At the Beigang Cast Iron Steelworks, awareness of the benefits of cleaner production had extended beyond management to the workers. The company had established a monitoring system for evaluating worker compliance and provided salary benefits and advanced enterprise awards to those workers who were actively involved in cleaner production activities. Non-compliance was being penalized with a reduction in salary.[99]

Despite such positive signals, project officials were concerned that they had not been able to set up a financial mechanism to ensure the continuation of efforts in cleaner production. Chinese banks were not willing to provide investment for such purposes. And the establishment of a revolving fund was not possible because the Benxi EPB was politically weak and, therefore, not able to collect the necessary fees from polluting enterprises.[100] Furthermore, it was not clear how the Benxi government was going to fund industrial restructuring. It was

Table 4.1 Industry benefits of the UNDP project in Benxi 1999

Factories	Economic benefits	Environmental benefits
Benxi Brewery	• RMB1.86 million from savings on resource inputs • RMB2.21 million from new production processes to recover and reuse waste products • reduced overall costs of production by 30 per cent	• Saved 255,000 tons of water, 139,000 kwh electricity, 46 tons of coal, and 80 tons of grain • Recovered 3,700 tons of waste liquid • Reduced chemical oxygen demand by 1,100 tons
Benxi Cement Factory	• RMB240,000 from mixing poor quality limestone with good quality limestone • RMB2.65 million overall profit (out of a total investment of RMB230,000)	• Reduced dust emissions by 2,000 tons (out of a total 5 million tons) further reduction of 8,000 tons of dust expected by end of 1999
Beigang Cast Iron Steelworks	• RMB500,000 profits (January–April 1999) out of a total investment of RMB2 million • RMB1 million profit expected by the end of 1999	• reduced solid waste by 688 tons, wastewater by 4,300 tons, and dust by 2,000 tons

Source: Based upon interviews and factory visits 18 May 1999.

Table 4.2 Achievements of cleaner production at Benxi Factory 2000

Item	Grains saved (per cent)	Water saved (per cent)	Coal saved (per cent)	Power saved (per cent)	Chemical oxygen demand reduced (per cent)
Target	1	5	2	2	10
Completion status	1.27	5.7	1	4.2	12.7

Source: www.chinacp.com/eng/cpfactories/cpfact_benxi_brew.html

abundantly clear that finding a substitute for Benxi's iron and steel industry – which in 1999 accounted for 37 per cent of local GDP but was placing a heavy burden upon diminishing natural resources – would require more than sound economic and environmental analysis.[101]

Mindful of the financial constraints involved in implementing the UNDP project, the director of the project office had decided to concentrate efforts upon public awareness building. Benxi is a classic example of how a crisis is needed before people are willing to take action. During the 1980s, the citizens of Benxi who had earned the dubious distinction of 'human vacuum cleaners' became actively engaged in voicing their concerns.[102] Ten years later, in the face of industrial restructuring, social welfare became the overriding concern (in 1999 there were 86,000 laid off workers in Benxi out of a total working population of 500,000) and therefore managers of industrial enterprises felt justified in placing profits before environmental protection. Consequently, local project officials believed that public-awareness raising was an essential prerequisite for both project success and future financial investment. In the words of the director of the project office:

> Cleaner production in the Agenda 21 Office is a wide concept. It encompasses not only technical change and management reform but also awareness building which is of crucial importance if our efforts are to succeed. To improve management awareness we cannot simply issue directives. Instead we need to work closely with companies to understand and resolve impediments to implementing cleaner production techniques.[103]

In November 1998 the Agenda 21 office organized a rally in the centre of Benxi to disseminate cleaner production information – 100,000 people attended. A year later an indoor meeting was convened for the heads of local government agencies with support from the city mayor. By 1999 over 2,400 local people had been trained in the concepts of sustainable development and cleaner production. Perhaps most importantly, in 2000, Madam Tang Guimei established a new NGO, the Benxi Sustainable Development Promotion Association, to promote the sustainable use of resources in Benxi and its surrounding mountainous areas on the basis of local participation. In relation to these initiatives, UNDP environmental assistance to Benxi had extended far beyond its initial project expectations with the promise of continued local support.

Capacity development for acid rain and SO₂ pollution control in Guiyang

Capacity development for acid rain and
SO$_2$ pollution control in Guiyang

The environmental project in Guiyang makes an interesting comparison because it was the first attempt by the UNDP to support capacity building in one of China's poorest provinces. At a total cost of just under US\$1 million (with the UNDP providing US\$609,000 and the Guiyang government a further US\$360,000) the project was primarily concerned with developing the city's monitoring and financial capabilities in the control of SO$_2$ emissions and acid rain. The project had four central objectives framed as follows:

1 To develop an economically sound SO$_2$ strategy for Guiyang.
2 To strengthen the capacity of the Guiyang EPB in decision-making and management for controlling acid rain.

3   To establish a mechanism for financing the necessary level of SO_2 control.
4   To lay the foundation for a local pollution control technology industry.[104]

At the time of my visit to Guiyang in June 1999 (two years after the launch of the project in May 1997) only the first objective seemed likely to be realized. With considerable technical support from CRAES in Beijing, the project office had succeeded in completing an emissions inventory and the field data was being used to develop an impact assessment and prediction model. This was no small achievement given that the city's expansion of its administrative boundaries in 1996 had meant that numerous sources of SO_2 were not being accounted for in existing models. With the expanded boundaries, SO_2 emissions had increased from 30,000 tons per annum to 500,000 tons.[105]

Guiyang also needed to improve its monitoring capacity before it could establish a more effective system of regulatory control.[106] Indeed, on the basis of the emissions inventory the local authorities had subsequently closed down a number of highly polluting units of the Guiyang Power Plant. However, very little progress had been made towards developing a cost-effective approach to controlling SO_2 emissions per se. Some control technologies had been evaluated which had relevance for the Japanese environmental loan project discussed in Chapter 3. For example, wet scrubbers had been identified as the most suitable technology for desulphurization in Guiyang. But overall, the project office was experiencing considerable difficulty in linking the technical dimension of pollution control with the economic dimension.[107] As in previous programmes to develop acid rain technologies sponsored by SEPA, the information required for the economic assessment of factors such as the operating cost had not been obtained. This was also mentioned in the UNDP evaluation report which recommended that the Guiyang EPB should focus its attention more on building the city's environmental equipment industry.[108]

The inability of local project officials to progress beyond a technical approach to pollution control can be explained by four factors. First, the project office was situated in the Acid Rain Center of the Guiyang EPB, the sole purpose of which was to improve monitoring capacity. Second, institutional arrangements were less than satisfactory. Although a leading group for the project had been set up under the chairmanship of the vice-mayor, the director of CRAES rather than the director of the Guiyang EPB was appointed as the national project director. This, of course, reflected a lack of confidence in the capacity of the EPB to implement the project. But the distance between the two cities (i.e. Beijing and Guiyang) meant that communication was minimal. The UNDP had no contact with the Guiyang project office whatsoever. Furthermore, the local project director, Madam Cheng Qun, did not share the same energy and enthusiasm as her counterpart in Benxi, Madam Tang Guimei; neither did she have the equivalent high political status.

Third, the intellectual infrastructure in Guiyang was weak in comparison to Benxi. The Guizhou University of Technology (formerly the University of Industry) and Guizhou University could not compete with Liaoning University either in environmental science or economics. The Benxi project had also benefited considerably from the economic input provided by the Centre for Environmental Sciences at Peking University,[109] whereas the Guiyang project was totally reliant upon CRAES, which is solely science-based. This seems to suggest that intellectual support is most needed in relation to bridging the divide between economics and environmental science. Fourth, and perhaps most important, local officials involved in the project lacked a clear understanding of the critical importance of building capacity for financial management as opposed to simply relying upon external assistance. One project official concluded:

> We have not been able to establish a financial mechanism for pollution control because we do not have the financial resources. Neither the local government nor enterprises are willing to pay for environmental protection. We need to attract more foreign investment such as the Japanese environmental loan project.[110]

It is interesting that both CICETE and CRAES in Beijing judged the UNDP project in Guiyang as a success in that it had helped to attract US$50 million in financial assistance from Japan.[111] The UNDP evaluation report also stressed that the project staff were convinced that the Sino-Japanese loan project would not have been awarded without the help of UNDP assistance.[112] But in reality, the Japanese chose to invest in Guiyang ostensibly for the same reason as the UNDP: because it is one of the worst acid rain areas in China. The UNDP project was not mentioned in any of the official documentation relating to the Japanese environmental project in Guiyang. Moreover, according to the Japanese project office in Guiyang, coordination between the two projects did not exist apart from the fact that the UNDP-funded SO_2 control plan had helped the Japanese to target the most polluting industries in Guiyang. Improved monitoring capacity also meant that the Japanese could more effectively monitor the impact of their environmental project. This benefit, however, did not appear to be the prime motivation behind Japanese investment in Guiyang.

The fundamental problem with the UNDP project did not lie in a lack of donor coordination but in the overriding belief at the local level that solving environmental problems depended more upon foreign investment than local capacity building. From a broader perspective, the failure to reconcile the technical and economic dimensions of the UNDP project in Guiyang did not bode well for the city's future success in integrating environmental concerns into economic planning and development.

Small is beautiful, or small is not enough?

In his seminal work during the 1970s, Friedrich Schumacher observed that small operations could lead to large impacts. He argued that large development projects were not suitable because of the unknown risks involved in large-scale technological advancement. In his words: 'There is wisdom in smallness if only on account of the smallness and patchiness of human knowledge, which relies on experiment far more than understanding.'[113]

This case study has demonstrated that under certain conditions small investments can indeed make a large impact. The UNDP human development approach to environmental management has led to tangible improvements in local institutional, technological and informational capacities. In Shenyang, the implementation of stakeholder participation has strengthened interagency coordination. And the introduction of cleaner production techniques in Benxi has enhanced the ability of some factories to manage the critical trade-off between environmental protection and economic competitiveness. This has, in turn, encouraged the Benxi municipal government to adopt ISO 14000 standards (the first municipal government to do so in China) in addition to setting up a Cleaner Production Centre within the EPB.[114] UNDP assistance has also helped to facilitate environmental information management in the cities of Wuhan and Guiyang.

From a broader perspective, the UNDP approach has had far reaching ramifications insofar as local perceptions are concerned. Above all, the approach has strengthened environmental capacity by encouraging a sense of shared responsibility. In the cities of Shenyang and Benxi the local EPBs are no longer seen as the only institutions responsible for environmental improvement; other government agencies and industrial enterprises have also demonstrated a willingness to share some of the burden. At the same time, public responsibility has been strengthened in Benxi through the establishment of an environmental NGO.

Yet the effectiveness of UNDP environmental assistance clearly varies. A human development approach relies upon intellectual support and public disclosure. Above all, the key determinant across all projects appears to be the *degree of pre-existing political commitment and institutional capacity*. Of course, political commitment is an elusive term and difficult to measure, but in the context of UNDP environmental assistance in China it is clearly commensurate with political status – both the status of the leaders involved in the project and the status of the implementing agency.

The UNDP's influence in Shenyang was not replicated in Wuhan for the simple reason that the necessary political arrangements were not in place: the mayor was less actively involved, the director of the project had a lower political status, and the EPB was relatively weak. Consequently, the project manager was in an untenable position. It would seem that the reason why the political will existed in Shenyang and not in Wuhan was related to intrinsic factors such as pride and status. Essentially, Shenyang had a strong political incentive to improve its environment because it had been singled out by the WHO as an international pariah – 1 of the 10 most polluted cities in the world. Environmental problems in Wuhan, on the other hand, had received far less attention both internationally and domestically and, therefore, were of less political consequence.

The contrast in political arrangements is no less striking when comparing UNDP environmental assistance in Benxi and Guiyang. In Benxi the UNDP project succeeded in demonstrating the importance of cost-effective technological innovation and in raising public awareness of the need for cleaner production – first and foremost because of the political leadership and status of the Project Director, Madam Tang Guimei. Under the Municipal Office for General Affairs, the Agenda 21 Office in Benxi also had a sufficient degree of political leverage.[115]

But in Guiyang, the implementing agency was politically weak and the project director was too far removed from local concerns, which seriously undermined the ability of the project office to wield the necessary local political and public support. The lack of leadership commitment was also noted in the evaluation report prepared by the UNDP in China in 2001. It states that 'the significant benefits of this project might have been even greater if the enthusiastic support of the EPB management was present throughout the whole project'.[116] The irony, therefore, is that in order to strengthen local capacity, a certain level of capacity needs to exist beforehand. Hence, where the approach is most needed – in the hinterland with limited foreign investment potential and weak institutional capacity – it is the least successful.

Even under the right conditions the approach is limited. The fundamental weakness of UNDP environmental assistance lies in financial management or the failure to establish a cost-effective means of ensuring the continuity of environmental projects. In the words of one official at the ACCA21 office in Beijing:

> The problem with UNDP environmental assistance is that it needs to address the issue of how to satisfy local needs for more physical infrastructure as opposed to simply capacity building, which is not useful if there are no funds to carry out recommendations.[117]

With small budgets, all of the implementing agencies were experiencing difficulties in establishing a financial mechanism to fund concrete plans of action which, in turn, had important implications for capacity building.[118] Without adequate capital, improvements in environmental capacity were not likely to endure over time.

Conclusion

This case study has demonstrated that a human development approach can lead to improvements in local environmental capacity in China. Overall, stakeholder participation and technological innovation have helped to strengthen local institutional and technological capacities. Moreover, the UNDP's environmental assistance to China has enhanced awareness of the need to share responsibility for environmental management, especially amongst government agencies and industrial enterprises.

Contrary to some of the expectations outlined in Chapter 2 of this book, the Chinese government at both the national and local levels is not totally resistant to the idea of participation. The Chinese people do have some say in environmental management, albeit indirectly through complaints hotlines, opinion surveys or the Internet. But the Chinese perception of participation is still elitist and, therefore, the UNDP has not lived up to its own expectations in relation to stimulating public participation.

Yet, it is also important to recognize that facilitating interagency participation in China is no small feat. It involves important changes in the local mindset, as illustrated in the following remark made by the Habitat advisor to the UNDP project in Shenyang:

> At first the concept of stakeholder participation was difficult to communicate in Chinese – they thought we were talking about driving stakes into the ground. The leading groups for the project were initially made up only of retired high level officials, which had to change. Also the Chinese could not see the logic in setting up working groups on strategic issues. After two years, however, there now exists a clear consensus that the working groups will be routinized into everyday practice.[119]

The investigation into the implementation of the same UNDP project in Wuhan clearly reveals the difficulties involved in implementing interagency participation in China. Again, in the words of the Habitat advisor to the project: 'Hu Qixi [project manager] is consistently walking a fine line between innovation in the decision-making process and allowing high status officials to pursue their vested interests.'[120]

In contrast, the implementation of technological innovation in China has been less fraught with political sensitivity. Chinese officials, especially at the local level, are becoming increasingly aware of the importance of introducing cleaner production and innovative methods of reducing pollution loads as opposed to simply relying upon expensive 'end-of-pipe' solutions. The issue now is how to ensure the diffusion of technological innovation throughout China. To this end, attention needs to be given not only to information dissemination, but also to seeking out innovative financial mechanisms to ensure that cleaner production is economically viable in the long term.

Overall, the main limitation with the UNDP's approach to environmental management is the weak linkage between capacity building and financial efficiency. For how long can local officials invest time and effort in capacity building when sufficient funding is not available to implement concrete plans of action? Despite differing degrees of success, the Shenyang, Wuhan, Benxi and Guiyang UNDP projects were similar in one important respect: local officials at all the project sites placed greater importance upon concrete outcomes rather than the process of environmental management per se. Herein lay the essence of the local response to UNDP environmental assistance to China: that strengthening managerial capacity was not enough; it was a means to an end rather than an end in itself.

Photograph 1 Living on the banks of the polluted river in Liuzhou.
Source: Taken by author.

Photograph 2 Factories belch out smoke in Benxi.
Source: Taken by author.

Photograph 3 Women protect themselves against air pollution in Shenyang.
Source: Taken by author.

Photograph 4 Selling wares at the edge of the garbage piles in Kunming.
Source: Taken by author.

Photograph 5 Wulong township in Yunnan province.
Source: Taken by author.

Photograph 6 Huang Long (Liuzhou Chemical Fertilizer Factory).
Source: Taken by Author.

5 Creating incentives and institutions

The World Bank approach

The World Bank's approach to environmental management in China is essentially market-driven with a continuing emphasis upon building physical infrastructure. It is to some extent guided by the results of the Bank's expanding policy and analytical work on environmental economics. With an equal emphasis upon establishing appropriate pricing policies and creating market-oriented institutions to operate and maintain physical infrastructure, the approach can best be defined as *market-institutional*. The underlying assumption is that securing an adequate economic return on investment will inevitably lead to positive effects upon institutional capacity. Indeed, for the Bank, developing capacity is largely synonymous with institution building. It has a fairly narrow conception of institutional building for environmental purposes by focusing upon the creation and development of market efficient environmental services. The Bank's position on capacity building, therefore, falls somewhere in between the minimalist approach of the Japanese and the targeted approach of the UNDP.

It would be misleading, however, to convey the impression that the Bank is exclusively interested in pursuing market solutions to environmental problems in China. In a recent departure from its usual commitment, the Bank has also been experimenting with participatory practices. In one environmental project in Guangxi Zhuang Autonomous Region, it has promoted a community-driven approach to environmental management in recognition of the fact that environmental problems in poorer urban settings require different economic and institutional arrangements. The latter is only a small component of a much larger project but it demonstrates a willingness on the part of the Bank to embrace alternative methods of aid delivery.[1]

One of the more surprising findings of this case study is that the Bank has been relatively more successful than the UNDP in promoting public participation, largely because it has taken a bottom-up approach. This is somewhat incongruous given that the Bank's comparative advantage lies in large-scale infrastructure projects. And it therefore calls into question whether an emphasis upon getting the prices right is the most important contribution that the Bank can make towards building environmental capacity in China, or whether a dual focus upon incentives and participation is more likely to be effective.

In the past the Bank's large-scale environmental projects have been heavily criticized for their lack of attention to local ecological and social factors. Yet, overall the findings in this study provide some grounds for optimism. The Bank's environmental projects in China, at least in the urban sector, are not an unmitigated disaster brought on by ill-conceived and hastily prepared plans drawn up in Washington. On the contrary, the projects are based upon a comprehensive analysis of local environmental needs and they have been relatively successful in strengthening local financial capacity. The market-driven approach has been less effective in dealing with environmental problems that reach beyond urban boundaries and involve multiple sources of pollution. Moreover, the Bank has encountered major difficulties in trying to build environmental infrastructure while, at the same time, creating the necessary market-oriented institutions for operating and maintaining it. These problems highlight the need for greater institutional flexibility both on the part of the donor and the recipient.

In order to explore the apparent contradiction in the Bank's poor track record in environmental protection and its recent practices in China, I will begin this chapter with a discussion of the broader institutional changes that have taken place at the World Bank since the early 1990s. This is followed by a discussion of the Bank's approach to environmental management, its core interests in providing environmental assistance to China, and the contentious issue of its policy influence at the national level. I then examine the local implementation of three World Bank environmental projects in the provinces of Liaoning, Yunnan and Guangxi Zhuang Autonomous Region. As in the previous case studies, the chapter concludes by assessing the donor's contribution to improving local environmental capacity.

The World Bank: change or continuity?

The World Bank Group consists of the International Bank for Reconstruction and Development (IBRD) founded in 1944, and its four affiliates – the International Finance Corporation (IFC), the International Development Association (IDA), the International Centre for Settlement of Investment Disputes (ICSID), and the Multilateral Investment Guarantee Agency (MIGA).[2] The IBRD and the IDA (hereafter referred to as the Bank) lend to governments and government-owned agencies and corporations.[3] They are the focus of this case study. The IBRD provides non-concessional development loans, with annual disbursements of around US$19 billion, funded primarily by borrowing on the international capital markets. Interest is charged at only a fraction of a per cent above the cost of borrowing – usually around 5.6 per cent. The IDA provides interest-free credits to low-income countries with 35–40-year maturities.[4] Like the UNDP, it is reliant upon voluntary contributions from governments. Annual disbursements are between $5 billion and $6 billion – almost three times the size of the UNDP's budget. The Bank employs over 6,000 professional staff and has representative offices supporting economic development projects and programmes in more than 90 countries.[5] By virtue of its financial largesse, together with its strong intellectual capital and over a half a century of development experience, it is no exaggeration to say that the Bank plays a pivotal leadership role in global development.

Yet, in the view of many environmentalists and development experts, the Bank's leadership role in regard to the environment has been pitiful at best, and highly pernicious at worst. Since the 1970s many scholars have written penetrating critiques of the Bank's failure to deal with environmental problems which provide a powerful image of a development Bank that is very much part of the problem rather than the solution.[6] This image, however, fails to capture the positive changes that have taken place in response to external pressure and institutional learning. Any objective assessment of the World Bank has to take into account change over time. For such a complex organization this is no easy task. The problem is compounded by the fact that most of the available literature written on the Bank is produced internally, which makes it difficult to distinguish between rhetoric and genuine commitment. Volumes have been written about the World Bank and it is not the intention here to review the major debates, but rather to focus briefly on some of the more recent changes that have led to the Bank's emerging role as a leading environmental donor.

The two faces of the World Bank

To all outward appearances, the World Bank has changed dramatically since its inception at an international conference at Bretton Woods in New Hampshire in 1944. At that time, the Bank's prime task was to help reconstruct post-war Europe. Loans were used to finance 'specific projects' and its earliest operations were primarily expected to facilitate foreign private capital flows. Attention turned to developing countries in the 1950s when it was realized that the Bank lacked the financial capacity for reconstruction in Europe.[7] Moreover, with the onset of the Cold War, the Bank had a strong political motive to channel resources to the Third World.

Over the years the development orientation of the Bank has shifted from purely infrastructure building to integrated projects targeting agriculture, oil and gas, urban services, water supply and sanitation, education, health and the environment. The Bank's programme is no longer solely focused upon projects but also includes structural and sectoral adjustment.[8] In addition, the Bank's research capability has been considerably strengthened since the McNamara administration in the 1970s. By the early 1990s, the Bank was employing over 800 professional economists with a research budget of around US$25 million per annum – considerably larger than the average annual budget of development institutes within academia.[9]

In tandem with these structural shifts, the development philosophy of the Bank has changed dramatically from 'trickle down' development in the 1960s, to export driven growth in the 1970s and 1980s, and to the current comprehensive development framework. The latter is based upon the rather belated realization that for economic development to take place it is necessary to pay equal attention to social, environmental, structural, human and governance factors. Key principles include a long-term comprehensive vision, ownership by the country, partnership with internal and external actors, and a focus on development outcomes.

In departing from the 'one size fits all' development doctrine of the past, the Bank is now more willing to take into consideration local needs and circumstances when drawing up development strategies and action plans. This is in part because, under the administration of James Wolfensohn, the Bank became more accountable for its past failings and, therefore, more open to experimenting with alternative development strategies. Consequently, the issue of participation is now firmly on the agenda. According to Bank officials, the impetus is coming more from middle management rather than from above. By 1999 an estimated 25 per cent of task managers (project managers) were actively using participatory methods in recognition of the fact that projects related to health, education, and water and sanitation (to name but a few) were highly likely to fail without the involvement of the beneficiaries.[10]

The shift towards comprehensive development has been further bolstered by the Bank's renewed commitment to global poverty reduction and the International Development Goals outlined at the Millennium Summit in 2000. To meet these goals, the Bank has become a leading advocate of the principle of 'development partnerships' between donors, NGOs and the private sector.[11] In aligning itself to the wider goals of the international community, the World Bank's development agenda has considerably expanded and is now little different from that of the UNDP.[12]

Yet, in other more fundamental ways, over the past half-century or more the Bank has changed very little. The Bank's original Articles of Agreement remain the same. Hence, the Bank's mandate is still to transfer resources to developing countries for productive purposes in pursuit of economic growth. The Bank's legal documents make no mention of the environment. Loans are to be used 'with due attention to considerations of economy and efficiency and without regard to other non-economic influences or considerations' (Article III, section 5 (b)). Moreover, 'except in special circumstances', the Bank is still required to lend for 'specific projects' (Article III, section 4 (vii)). The prevailing orthodoxy, at least in strictly legal terms, continues to be one of growth, albeit sustainable growth where possible.

Basic infrastructure continues to be important (although the share of total lending declined to around 28 per cent in 2003)[13] and lending is still concentrated on a small number of large projects (the average loan is approximately US$50 million for a total project investment of US$140 million). This is largely a reflection of the Bank's lending imperative. In order to maintain its status as a net provider of funds to developing countries, the Bank must continue to increase its lending or otherwise face the threat of liquidation. As interest accumulates on earlier borrowing, the Bank faces the real risk of net negative transfers – when borrowers are paying more in interest and principal to the Bank than they are receiving in loans.[14] As Edward Mason and Robert Asher note, the Bank cannot be called a development institute if it is collecting more from the poor than it is lending.[15]

The problem is that the career incentives for project managers working at the Bank lie in disbursing large sums of money rather than in actual project performance. Traditionally, quantity has far exceeded quality.[16] Moreover, the Bank is under no external competitive pressure to take the quality of project outcomes into account. Like SOEs in China, the Bank faces a soft-budget constraint – its loans are guaranteed and subsidized by national governments and do not entail any financial risks. Approximately 7 per cent of IBRD's subscribed capital is paid in, with the remaining 'callable capital' (US$175 billion in 1998) acting as a guarantee in support of the Bank's borrowing from the international capital market.[17] It may be used only in the event that IBRD is unable to meet its obligations towards its creditors. But in practice this has never happened. No borrowing country has ever defaulted on a loan because the Bank enjoys a 'preferred creditor status' (developing countries repay debts from new loans and the World Bank takes priority because it is generally understood amongst developing countries that to renege on a World Bank loan would seriously undermine their credit status).[18]

The two sides of the Bank can be likened to the Roman God Janus – the lighter side is responsive to new ideas and practices and the darker side is trapped in fiduciary concerns. They are a constant reminder that the Bank is both a development and a financial institution. As Mick Moore argues: 'The Bank, financed through borrowing on the world's capital markets, could not adopt the same "welfarist" stance as organisations such as the UNDP and UNICEF that are grant funded.'[19] Project lending must contribute to the development of a borrowing country and at the same time to its capacity to repay the loan. In addition, no matter how progressive on paper, the Bank's development activities are still circumscribed by its dependence upon the world's richest industrialized countries. Member subscriptions to the Bank's share capital determine voting rights. Amongst the Bank's Executive Board of Directors, the United States continues to hold a large share of voting power (although this has declined over time from 25 per cent in the 1970s to 15 per cent in 2003)[20] and often uses it for self-serving ends.[21]

When a new policy objective such as environmental protection fails it is not so much for lack of trying as the fact that very often development policy goals run counter to the *modus operandi* of the Bank. For example, it is difficult to integrate environmental concerns into Bank lending when staff receive little reward for doing so. Also it is not possible to incorporate participatory practices into Bank projects when the system puts greater emphasis upon project preparation rather than implementation, and when the Bank is under constant pressure to disburse large sums of money. Not surprisingly, and notwithstanding the observations above, genuine efforts to integrate environmental concerns into the Bank's operational activities have proven to be highly problematic. The Bank has tried to address this issue by introducing Green Awards. However, it is too early to tell whether this incentive is having any real effect on actual practice.

The issue of 'mainstreaming' the environment (in Bank parlance) has been the subject of many internal studies and scholarly works.[22] Considerably less attention has been given to the Bank's widening portfolio of environmental assistance. As Robert Wade notes, the Bank's environmental projects are more likely to be effective because they are sectorally based and 'fully consistent with the Bank's long-established mode of organisation'.[23] It would be misleading, however, to discuss the Bank's environmental assistance in isolation because it is inextricably linked to the wider agenda of integrating environmental concerns into Bank practice.

The evolution of World Bank environmental lending

For the Bank, environmental lending is important because it provides institutional legitimacy. Although internal factors have been involved in shaping the Bank's environmental lending, it is fair to say that external pressure has been a central driving force. The rising tide of criticism of the Bank's activities by environmental activists during the 1980s and 1990s has left an indelible mark on the institution's environmental record. The environment first became an issue for the Bank in the early 1970s, when it played a leading role in advocating the importance of environment-development linkages at the UN Conference on the Human Environment held in Stockholm in 1972. At the time, the Bank was seen to be an environmental leader amongst bilateral and multilateral development institutions simply by virtue of the fact that, as early as 1970, it had set up an environment office to evaluate all Bank projects.[24] The Bank's environmental agenda, however, was hardly progressive; it was narrowly based upon a 'do-no-harm' philosophy that stressed environmental safeguards such as environmental impact assessments with little attention given to the linkages between the environment and development.

Substantial environmental reforms did not take place until the late 1980s when the environment became a major public relations issue for the Bank following the public outcry over numerous cases of Bank-funded projects that had caused egregious damage to the environment.[25] Ironically, some of the worst cases such as the 1981 Polonoreste project (a road building and agricultural colonization scheme to support settlers in the export of coffee and cocoa in Brazil's northwest) were first conceived of as examples of environmentally sound investments.[26] In actuality, Polonoreste was badly planned from the outset. The institutional arrangements were not in place at the time of implementation: land titles had not been secured, and the necessary credit services to support the settlers had not been set up. Consequently, thousands of settlers arrived and for their survival had little choice but to turn to burning forests in order to grow rice and maize. According to Bruce Rich, deforestation increased from 1.7 per cent in 1978 to 16 per cent in 1991.[27] The social consequences were equally dismal – settlers and the indigenous population suffered badly from malaria and other life-threatening diseases.

In regard to the Bank's subsequent efforts to make amends for such debâcles, two important points need to be made. First, projects such as Polonoreste, the Indonesian transmigration scheme, or the Singrauli energy project in India (to name but a few) were not only unmitigated environmental disasters; they were also a failure economically. With hindsight, recognition of this fact undoubtedly had an influence upon the Bank's later efforts to link environmental protection or improvement with economic efficiency. Second, the Bank's environmental disasters were also social disasters. NGO-led critics of the Bank were not solely focused upon the environment but also upon the issue of social equity. Hence, in the 1990s, in response to this criticism, the Bank adopted a broad conception of the 'environment' (i.e. socially and environmentally sustainable development) that encompassed issues of resettlement and the rights of indigenous peoples.

Yet, the Bank's negligence with respect to the environment was not formally acknowledged until 1987 under the presidency of Barber Conable, and only then, according to Wade, because Washington-based NGOs had effectively lobbied the United States Congress which, in turn, was threatening to withhold its contribution to the IDA.[28] The Bank announced major reforms in May 1987 including the establishment of an Environment Department and National Environmental Action Plans (NEAPs). The latter, as Wade notes, were instrumental in shifting concerns away from 'end-of-pipe' solutions and towards the comprehensive integration of environment-development linkages.[29]

The NEAPs triggered the beginnings of the Bank's 'environmental' projects, which were at first mainly focused upon forests and agriculture. They were criticized by some commentators as being counterproductive.[30] But during the 1990s the Bank made substantial progress in its environmental lending programme. Above all, major breakthroughs were made in relation to energy and forests. Supply-side energy lending declined over the 1990s from US$5 billion in 1992 to US$2.15 billion in 1998.[31] At the same time, lending for energy conservation (renewables and energy efficiency) increased. This is, in part, a consequence of the 'leveraging effect' of GEF funds mentioned in the UNDP case study. During the 1990s, as noted by Charles Feinstein *et al.*, each dollar of GEF financing attracted approximately US$1.25 in associated Bank financing.[32]

With respect to forests, the Bank is now placing a high priority upon 'community-based' programmes to enhance forestry management. Towards this end, the Bank has a strong commitment to work in partnership with NGOs as in the case of the World Bank–World Wide Fund for Nature (WWF) alliance for forest conservation and sustainable use.[33] Obviously, given the crucial role of forests in influencing carbon cycles, the Bank's growing concern with forests is in accordance with its international support for reducing the output of greenhouse gases. By 1998, the Bank's forests portfolio totalled US$4 billion (including IBRD/IDA and GEF funds).[34]

Total environmental project lending increased dramatically during the 1990s from US$1 billion in 1991 to US$7 billion in 2000.[35] Pollution control dominated the Bank's environment portfolio, especially in Asia and Latin America. Between 1988 and 1998, over half of the Bank's cumulative environmental lending (US$13 billion in total) was allocated to pollution and urban environmental management[36] (US$7.2 billion), with the remaining allocated to natural resource management (US$4.3 billion), GEF-funded global issues (US$1 billion), and institutional development (US$0.5 billion) (see Figure 5.1). By 1998, the world's four top polluting countries – China, India, Mexico and Brazil – were receiving approximately 40 per cent of the Bank's environmental funds.[37] More recently, the sectoral composition of the Bank's global environment portfolio has changed in favour of greener issues. Natural resource management constituted 46 per cent of total

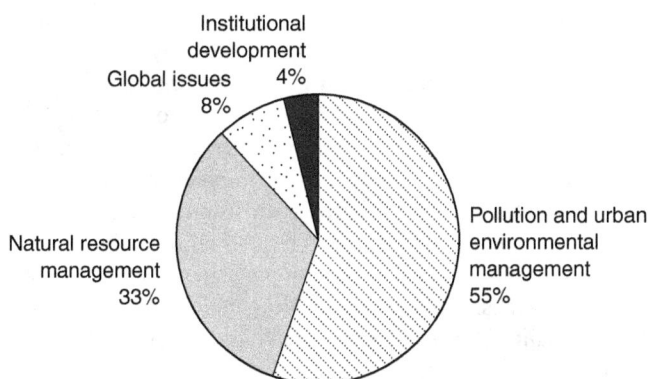

Figure 5.1 The sectoral composition of World Bank environmental lending 1988–1998.

Source: Calculated on the basis of the World Bank, 'Environmental projects portfolio', in *Environment Matters at the World Bank: Annual Review*, Washington, DC: World Bank, Fall 1998.

environmental funds in 2001. However, in the East Asia and Pacific region pollution control projects remain dominant (constituting approximately 57 per cent of the region's environmental funds in the same year).[38]

The Bank's projects are labelled as 'environmental' 'if either the costs of environmental protection measures or the environmental benefits accruing from the project exceed 50 per cent of total costs or benefits'.[39] But the Bank has found it difficult to define what exactly constitutes an environmental project. In the past, a large number of projects that could have been classified as environmental, relating to agriculture, rural water resources and some industrial projects, were excluded from the Bank's environmental lending profile. This situation has now been rectified with the introduction of a new methodology that takes into account the extent to which environmental concerns are being mainstreamed into the Bank's overall projects and programmes. Environmental lending now includes core environmental projects, other sectoral projects with primarily environmental objectives, and projects with minor environmental components. On this basis, total environmental lending in fiscal 2000 reached a peak of nearly US$18 billion, excluding US$1.5 billion allocated to GEF financing.[40] This later declined to US$14.4 billion in fiscal 2004.[41]

The Bank has come a long way since its 'eco-predator' days of the 1980s. It now has a vice-president for Environment and Socially Sustainable Development, an impressive environmental research base and cross-sectoral environmental, social, and rural knowledge networks. Noteworthy innovations include the development of a Global Carbon Fund through a partnership with the GEF to increase the development of renewable energies, the establishment of a World Commission on Dams to improve environmental and social standards through a partnership with the World Conservation Union, and the launch of a Clean Air Initiative to build capacity for improved urban air quality in Asian cities.

More recently, the Environment Department[42] has played a major role in developing a new environmental strategy for the Bank on the basis of a two-year consultation process involving other donors, NGOs, academics and policy-makers in recipient countries. This represents an important turning-point in the Bank's environmental thinking: it is no longer simply reacting to negative external pressures, but instead forming its own internally driven framework in collaboration with external actors. Published in 2001, the *Environment Strategy* aims to clearly outline how the Bank intends to address environmental problems in client countries and how it plans to integrate environmental objectives into the Bank's core operations. In linking environment with poverty reduction, the strategy sets out three main priorities: improving people's quality of life, improving the prospects for the quality of growth, and protecting the quality of the regional and global environmental commons.[43]

It is too early to tell how the new strategy is affecting actual Bank practices. Given the emphasis upon poverty–environment linkages, it is likely that future environmental projects will shift to the poorer areas and regions of the world. In the meantime, the Bank's approach to environmental management in urban areas remains deceptively narrow. This largely reflects the fact that the Bank's comparative strength in environmental protection and improvement still lies in improving economic efficiency.

Bringing in the market

Ultimately, economic considerations define the Bank's approach towards assisting developing countries in environmental management. Fundamental to the economic-centred model of environmental management is the belief that sound fiscal management – by pricing natural resources such as water and fossil fuels to capture the cost of their use – is critical to ensure the sustainability of environmental investment. For the Bank, the need for appropriate pricing incentives is perceived as universal, regardless of income differentials between regions and issues of social inequality. Consequently, the Bank has been criticized by some scholars for its 'north-biased accounting perspective'.[44] According to the Bank, this view is shortsighted for two important reasons. First, whereas discount rates amongst the poor with respect to conservation tend to be high, this is not the case for pollution; the poor are often willing to pay for clean water and sanitation. Second, the health costs from drinking polluted water or buying water from hawkers far outweigh any incremental costs imposed by improved environmental services.[45] This seems to suggest that a pricing strategy for environmental improvement would be effective in China despite its huge disparities in income.

The importance of *getting the prices right* is not new. The Bank stressed the need for public utility price reforms, with a particular emphasis upon removing agricultural subsidies, well before the environment became a serious issue for the Bank in 1987.[46] But the link between pricing incentives and environmental protection was not clearly established until the early 1990s. To this end, the 1992 *World Bank Development Report: Development and the Environment*, played an important catalytic role in linking the basic economic policy orientation of the Bank with desired environmental outcomes. The 'win-win' scenarios posited in the report, first and foremost, underlined the economic logic of protecting the environment, framed as follows:

> Good environmental polices are good economic policies and vice-versa. Efficient growth need not be an enemy of the environment, and the best policies for environmental protection will help, not hurt, economic development.[47]

In order to reconcile development with environmental protection the emphasis was upon removing subsidies and setting appropriate pricing policies for water, electricity and even timber. The Bank's report made no further progress in identifying exactly how local institutions in developing countries could manage the inevitable trade-offs between environmental protection on the one hand and economic efficiency and growth on the other. This was purely for pragmatic reasons. As Wade argues, 'if it had been more nuanced, more concerned with trade-offs, if it had admitted that on many questions the evidence is not clear, [the report] would have made less impact inside and outside the Bank.'[48] As it stood, the report not only helped to mollify the development concerns of its major clients but also had the effect of galvanizing support for environmental protection amongst the mainstream economists within the Bank. Not all economists were converted, however. Before the report was released a leaked memorandum by the Bank's then chief economist, Lawrence Summers, on 12 December 1991 gave the strong impression that the market analysis of environmental problems still dominated World Bank thinking:

> Just between you and me, shouldn't the World Bank be encouraging *more* migration of the dirty industries to the LDCs?...I think the economic logic behind dumping a load of toxic waste in the lowest-wage country is impeccable...I've always thought that under-populated countries in Africa are vastly *under*-polluted; their air quality is probably vastly inefficiently low [*sic* high] compared to Los Angeles or Mexico City.[49]

While the Bank's more critical work in ecological economics, advanced by Herman Daly and Robert Goodland from the Environment Department, remains largely divorced from the Bank's day-to-day activities, its expanding research agenda in environmental economics has been more successful in infiltrating the prevailing neo-classical economics approach to dealing with environmental externalities. This is because environmental economics does not stress the finite nature of the earth's resource base and hence, is more amenable to reconciling economic growth imperatives with ecological preservation. The Bank's applied and theoretical work in environmental economics has considerably strengthened the Bank's capacity to conduct economic analyses of environmental problems, in addition to having some influence over policy. By the mid-1990s the Bank's research establishment was advocating a new paradigm for environmental management which linked the state with the market and communities in a way not dissimilar to the Japanese model of environmental management in the 1960s. Four key policy approaches were delineated as relevant to both resource management issues and pollution: using markets (taxes and fees), creating markets (tradable permits and rights), regulation and engaging the public.[50]

But a large gap exists between the policy prescription advanced within the Bank's research establishment and the Bank's actual achievements on the ground. In practice, the Bank's approach to environmental management has remained centrally concerned with the first policy area – using markets – with some attention being given to the introduction of market-based instruments in its policy dialogue with recipient countries. Institutions are also considered to be important. For the Bank, institutional development with respect to the environment covers a broad spectrum of activities involving the decentralization of environmental policies, privatizing water supply, solid waste management and sewage collection etc., and promoting public–private relationships. The market-institutional approach to environmental management is particularly dominant in the urban sector where Bank studies across a number of regions have highlighted the human and economic costs of air and water pollution and hence, the need for pricing incentives and the provision of efficient urban services.[51]

The approach leaves little room for participatory practices. Although there exists a growing consensus within the Bank that participation is an important means of developing effective environmental protection (particularly in those countries where institutions are uniformly weak), it is not widely integrated into the Bank's current environment portfolio. Participation is mostly promoted in relation to forestry projects with limited spillover into urban environmental management. Participatory sanitation management in the *favelas* in Brazil and support for participatory-style environmental planning activities in Latin America and East and Central Europe are notable exceptions. Like the UNDP, the emphasis here is upon stakeholder participation defined in 1994 by the Bank's Learning Group on Participatory Development as follows: 'Participation is a process through which stakeholders influence and share control over development initiatives and the decisions and resources which affect them.'[52]

The Bank's dominant approach to environmental management, with its focus upon incentives and institutions, is deemed to be the most effective means of ensuring the sustainability of environmental investment while, at the same time, creating a positive externality in building local capacities for financial and institutional efficiency. The question is: to what extent can the approach apply to China with its enduring socialist institutional structures, huge income disparities across regions, and slow and haphazard market reforms? The Bank's position discussed earlier is that pricing incentives can be applied to developing countries regardless of income disparities. But the Bank has provided no evidence to suggest that this approach can equally apply to developing countries that are also in the process of transition from central planning to market socialism.

The greening of World Bank environmental lending to China

China became a member of the World Bank in May 1980. In the early years, lending programmes focused upon rebuilding the country's higher education system, which had been devastated after a decade of the Cultural Revolution, and thereafter concentrated upon supporting China's modernization drive and economic reforms.[53] China became the Bank's largest borrower in 1992, receiving roughly US$2.5 billion per annum (accounting for approximately 9 per cent of the Bank's overall lending). Total World Bank commitments to China as of June 2002 totalled US$33.9 billion involving 239 projects.[54] Lending by sector has overwhelmingly targeted infrastructure development (energy and transportation) and agriculture.

Cumulative lending for the environment (water supply, sanitation and urban development) has reached about 11 per cent of total funding to China. In recent years, the environment has become the fastest growing area of the Bank's operations in China. Approximately 25–30 per cent of the Bank's lending to China since the 1980s has been environment-related. In 1999, the proportion increased to 50 per cent with the decline in supply-side energy lending mentioned earlier.[55] Over the 1990s, the sectoral allocation of the Bank's environmental lending to China favoured pollution control and urban environmental management (total US$1.8 billion) over natural resources management (total US$0.7 billion). GEF and Montreal Protocol supported projects accounted for 10 per cent of total environmental lending and only a small amount targeted institutional development. Environmental assistance was largely in the form of IBRD loans. IDA credits to China were mainly used to assist poor regions in the northwest and southwest. The strong emphasis upon brown issues has continued with approximately US$1.4 billion disbursed for projects relating to pollution control and urban environmental management between 1999 and 2003.[56] At the same time, a greater emphasis has been placed upon dealing with climate change. This became the Bank's top environmental priority in China in 2003.

The World Bank's environmental assistance to China is based upon its 1992 *Environmental Strategy* (in two volumes) that was carried out in collaboration with government agencies in Beijing. The strategy proposed a multi-year programme consisting of three main areas: environmental sector studies (to provide the analysis and preparatory work necessary for the formulation of policy, institutional action plans and investment programmes), technical assistance (to strengthen environmental institutions) and environmental lending to support improvements in resource efficiency and the reduction of pollution. The report identified particular problems with China's economic incentive measures: pollution fees were too low in order to provide the necessary incentive for industries to combat pollution; and the responsibility system did not reward cost-effective ways of meeting pollution control goals.

A follow-up report on urban environmental service management in 1994 provided detailed recommendations for implementing an economic approach towards environmental reform – again with particular attention to pricing incentives. The report praised China's continuing reform of energy prices (in the more developed regions of China electricity prices are now similar to those in Australia) but urged that greater attention be paid to reforming prices for urban environmental services such as water supply and wastewater treatment. It concluded on a cautionary note:

> While the imposition of user fees may be politically difficult, city leaders must recognise that the costs will be paid directly or indirectly, and that the current choice to rely on indirect payments leads to excess demand, on the one hand, and on the inability to meet needs, on the other. The result in the end is greater pollution and higher total costs for the service provided.[57]

Three years later, the China 2020 Environment Report focused entirely upon the potential economic and social benefits of investing in pollution control as an intrinsic part of China's future economic development. The report proposed a new growth strategy for China based upon three key principles: harnessing the market to work for the environment, not against it, harnessing growth for the environment by pursuing investments with the highest environmental benefits for future generations, and harnessing China's administrative capabilities for the environment including wider community participation in environmental policy-making.[58]

Environmental assistance to China has largely evolved in keeping with the market-oriented sentiments of the above three reports.[59] The issue of participation was not addressed until very recently with the implementation of a participatory-style community sanitation project in Guangxi.[60] Small amounts of funding have been provided to develop the SEPA's capacity to implement environmental policies and action plans.[61] But by and large, the Bank's capacity building efforts have concentrated upon developing the necessary institutional support for environmental infrastructure in the form of market-oriented environmental services. In the 1990s, with the exception of the Loess Plateau Watershed Rehabilitation Project (at a cost of US$260 million), the Bank's large-scale environmental projects focused upon urban environmental management. More recently, attention has turned to natural resources management with the launch in 2002 of a new Sustainable Forestry Development Project (at a cost of US$214 million). This is in keeping with the new strategic priorities outlined in the Bank's 2001 environment report on China that stresses the need to tackle both pollution control and ecological degradation on the basis of improved institutional arrangements, new policy and economic instruments, and more efficient investments.[62]

World Bank motives

The World Bank's motives in providing environmental assistance to China are ostensibly pragmatic. First and foremost, China is an important client. As mentioned earlier, China has been the Bank's largest borrower since 1992. It is also the Bank's largest recipient of environmental funds – receiving US$2.9 billion out of a total of US$13 billion (almost triple the total amount of environmental lending to Africa) by 1999.[63] Staff at the East Asia Sustainable Environment Network, which is part of the Environment Department, openly acknowledge the fact that without China they would not have a programme.[64] But the real advantage in providing environmental assistance to China seems to lie in the scope for disbursing large sums of money (on average between US$100 million and US$150 million per project) over a relatively short period of time, thereby satisfying the Bank's lending imperative and meeting its development objectives at the same time.

A more cynical view is that the Bank is funding environmental projects in China in order to compensate for the environmental externalities of its regular operations and thereby obviate the need to integrate environmental concerns into all of its development projects. In the words of Korinna Horta, '[Environmental lending] undermines much-needed efforts to ensure that development projects internalise their social and environmental costs, a pre-condition of sustainability.'[65] However, this does not seem to be the case in China for two reasons. First, the Bank makes enormous efforts in its preparatory studies to ensure that lending actually meets local environmental priorities; and second, with the substitution of energy supply projects with energy conservation projects the overall negative impact of World Bank lending upon the environment in China has been considerably reduced.

A second motive is that China remains an important target of international environmental funds. Although the Bank does not suffer from the same budgetary constraints as the UNDP, it still stands to gain from attracting these funds because they can be 'leveraged' onto pre-existing Bank programmes. Moreover, from a purely institutional perspective, the availability of GEF or Montreal Protocol funds provides an additional incentive for the Bank's task managers to increase their number of environmental projects. By fiscal 2003 climate change-related projects accounted for 30 per cent of total environmental lending.[66]

Finally, China is an important research site for the World Bank. The Bank's impressive knowledge base on environmental issues has in large part evolved on the basis of its collaborative research in China – especially in relation to industry pollution and urban management. The *Clear Water and Blue Skies* report written by Todd Johnson *et al.* in 1997 has been widely used to illustrate the economic dimension of environmental problems, not only in China but also in the East Asia region as a whole. Hence, it is fair to say that without the Bank's involvement in addressing environmental problems in China, its research capacity on environmental management would be far more limited.

To summarize, the World Bank's interests in providing environmental assistance to China are both financial and developmental and, therefore, accord with the Bank's twin institutional imperatives. With the urban share of the population in China expected to reach 42 per cent by 2010,[67] it is little wonder that the Bank is keen to be involved in strengthening urban environmental management in China – particularly given the central government's dominant concern with alleviating water and air pollution. The Bank's approach to environmental management lays equal emphasis upon pricing incentives and market-oriented institutions to provide environmental services. The approach, at least in theory, is conducive to China's current commitment towards market reform. But given the continuing legacy of central planning in China, tensions between old and new institutional practices seem unavoidable.

Negotiating influence at the national level

The administration of the Bank's development activities in China is divided between Beijing and Washington. Although the World Bank representative in Beijing acts as the Bank's country director for China and is responsible for project approval, actual project management still depends upon task managers based at headquarters. Policy dialogue takes place between the Bank's representative office in Beijing and its counterpart agency, the MOF, with the support of Bank missions from Washington. The NDRC and SEPA are also involved in policy negotiations. The administration of World Bank projects is far simpler than in the cases of the UNDP and Japan because fewer agencies are involved and it is more streamlined. For example, the Ministry of Finance has a World Bank department with sub-offices in all the provinces and municipalities where Bank-funded projects are located.

Despite the relatively well-coordinated administration, the issue of the Bank's policy influence in China remains unclear. In a survey of the Bank's first half-century carried out by the Brookings Institution in 1997, the authors claimed that 'with very few exceptions, the intimacy and rapport between the Bank's Beijing resident mission and the Chinese authorities were unmatched in the Bank's history'.[68] The view expressed by the representative office in Beijing, however, was less sanguine. A former World Bank representative, Pieter Bottelier, asserted in 1996 that 'more than elsewhere they [the Government of China] have used us, and they have always been in the driving seat'.[69] Task managers currently working in China confirm that the Bank needs to have more policy influence in China, but they also admit that the situation is improving. In 1999, for example, the central government made an unprecedented request to the Bank for a policy note on urban development to be incorporated into its upcoming tenth Five-Year Plan (2001–2005).[70]

Despite these changes, it would seem that relative to other developing countries the Bank's leverage in China is limited as a consequence of three interrelated factors: China's size and importance in the world economy, its sensitivity to external interference and its huge borrowing potential. But this is not to suggest that the Bank has no influence over policy in China. In relation to the environment, for example, the Bank has played a major role in convincing the Chinese government of the need to take preventive action by stressing the costs (both economic and social) of not doing so. The important point to make here is that the Bank's credibility at the national level in China lies in the fact that its policy recommendations are grounded in empirical studies carried out on a collaborative basis with government agencies and research institutes in Beijing.[71] The Bank's extensive research studies on China's pollution charge system and regulatory compliance, for example, have been used during policy negotiations to show how China would benefit from higher pollution charges.[72] In the same manner, the Bank has been able to build a consensus in Beijing over the issue of utility price increases and the reform of environmental services. More recently, it has had an important influence over China's shift towards experimenting with economic instruments for environmental purposes.[73] As a consequence, the Bank has been able to play a significant role in supporting the implementation of the Clean Development Mechanism in China.[74]

Yet, the World Bank–China relationship is not free from political constraints. Tensions were particularly high in the lead up to the withdrawal of IDA funding to China on 1 June 1999.[75] In Chinese eyes, the decision was 'purely political'; it illustrated the dominant influence of the United States on World Bank operations. In the words of one SEPA official:

> We believe that it is some conspiracy led by the United States government. We have heard that the voice of the United States in World Bank lending operations is constantly blocking any attempt to improve funding to China. The reason is political because there exists no rational reason to withdraw IDA credits – China is still a developing country.[76]

At the time, in the absence of any reasonable motive, it was hardly surprising that agencies in Beijing were suspicious of the World Bank's intent. The IDA provides credits to poorer countries with a per capita GNP of less than US$925 – strictly speaking, in 1999 China had not yet reached this threshold. More importantly, China's income per capita continues to differ widely across regions – per capita income is four times greater in Guangdong than in Guizhou, for example. The withdrawal of IDA funding also makes little sense if the Bank intends to continue with its environmental and poverty alleviation projects in China. The Bank's environmental projects in poor regions of China have important IDA components. Moreover, the critical institutional development dimension of the projects is funded in part through IDA credits. Without these 'softer' components it is questionable whether the Bank's environmental loans will remain in demand in China. With respect to infrastructure development, Japanese environmental loans are more competitive (with a 0.7 per cent interest rate compared to the Bank's standard rate of 5.68 per cent).[77]

Aside from the politics surrounding IDA funding, tensions have also risen over the financial and accountability aspects of environmental lending. Unlike their counterparts in Tokyo, task managers in Washington are acutely aware of the difficulties involved in providing environmental funds to non-performing industrial enterprises that might possibly be facing imminent closure on economic grounds. Furthermore, in the absence of a reliable banking system, they are uncomfortable with having to depend upon SEPA to channel financial resources effectively.[78]

The real problems for the Bank, however, lie at the local level rather than the national level. For the Bank, the question is not so much how to influence Beijing but how to influence provinces, municipalities and counties where the Bank's projects are actually being implemented. Above all, at the local level the Bank has experienced numerous difficulties in implementing tariff adjustments. In

Shanghai, for example, despite assurances to the contrary, local agencies refused to implement a water pricing policy, which they perceived to be unpopular. In response, the Bank withdrew funding. Following a long delay, the Shanghai government agreed to implement the policy only after a telephone call from Zhu Rongji. The former premier has a strong political constituency in Shanghai and, therefore, the policy was implemented straight away. In Hubei province, the Bank experienced a similar problem. An urban environmental project was delayed for nine months because, despite assurances to the contrary, pricing reform was not being implemented.[79] Consequently, rather than relying upon trust, the Bank now requires to see actual bills as proof of implementation.

While these confrontations were taking place, the Bank's resident representatives in Beijing were making unsubstantiated claims with respect to the high success rate of the Bank's operations in China. Former acting resident representative Austin Hu maintained in 1997 that 'the success rate for project implementation is more than 95 per cent in China, well above the world average of around 70 per cent'.[80] A year later, Yukun Huang, the incumbent World Bank representative confirmed that 'the Bank's portfolio of projects [in China] continue to be one of the best in the Bank'.[81]

Perhaps not surprisingly, given the high political and economic stakes involved for both donor and recipient, there appears to be a huge disparity between the Bank's rhetoric in Beijing and what is actually happening on the ground in China that reinforces the need for investigations at the local level. The discussion that follows examines the implementation of three World Bank-funded urban environmental projects. These projects are designed to improve urban environmental problems in a comprehensive way by integrating municipal infrastructure building with industrial pollution control, policy reform and institutional development. In keeping with the central thrust of the Bank's approach to environmental management, the analysis will focus upon the policy and institutional aspects of the projects as opposed to the investments in physical infrastructure. Special attention will be given to identifying how the Bank's approach is affecting local environmental capacity, and under what conditions.

Implementing environmental projects at the local level

Throughout the 1990s, the Bank channelled nearly US$3.5 billion to improve urban water and air quality in China. In contrast to Japan, the Bank has a strong preference for dealing with water rather than air pollution control. The latter is considered to be problematic for three reasons. First, it is difficult to pinpoint responsibility because many sources are involved – households, power stations, transport, etc. Second, the necessary technology is relatively expensive. Third, air pollution control has cost-recovery problems because unlike the case of water supply or sewerage, it is not possible to establish environmental services and impose user fees in order to ensure a return on investment.[82] The Bank's thinking on this matter is illustrative of its overall predisposition towards combating environmental problems that have largely economic solutions.

To date, the Bank's urban environmental projects have been located mainly in the richer regions of China (Shanghai, Tianjin, Hebei, Beijing, Jiangsu and Zheijiang) with a more recent shift towards the poorer western regions (Yunnan, Guangxi and Sichuan). By 2004, 13 integrated urban environmental projects were under implementation. The three project sites in this case study have been selected to take into account regional variation: one project is located in the relatively developed region of Liaoning province and the other two projects are located in the poorer regions of Yunnan province and Guangxi Zhuang Autonomous Region. The projects are province based and involve cities, prefectures, counties and, in some cases, townships and villages.

The overall objective of all three projects is to develop environmental infrastructure and overcome the cost-recovery problem involved in the Bank's environmental investment by reforming prices and strengthening environmental services. All the projects are required to increase water tariffs and establish utilities – wastewater/drainage and solid waste treatment companies – that are financially autonomous.[83] The Bank provides, on average, 50 per cent of the foreign exchange component for the projects with counterpart funding from the beneficiaries involved. As of 2004, only the Liaoning project had reached completion.

The provincial or regional government at each location has established a leading group to oversee project implementation headed by the governor or vice-governor of the province. Provincial or regional implementation units, usually located within the environmental protection bureaux, were directly responsible for the day-to-day management of the projects. Sub-project offices have also been set up at each of the project cities. The projects share similar goals, institutional arrangements and levels of political commitment but the local responses to the projects vary. Despite differing levels of economic development between the project sites involved, variance in project effectiveness is largely determined by spatial rather than economic factors, according to the degree of urban concentration within the local setting.

Integrated environmental management in Liaoning province

The World Bank's urban environmental project in Liaoning was amongst the first cohort of Bank-funded projects in China that addressed urban environmental problems in a comprehensive way. The Bank's environmental involvement in Liaoning province dates back to the early 1990s, when it invested in the Liaoning Urban Infrastructure Project to strengthen water supply systems in the basin cities of Shenyang, Fuxin and Yinkou. For this purpose, a project implementation unit, the Liaoning Urban Construction and Renewal Project Office (LUCRPO), was established under the Provincial Construction Commission in Shenyang. Following this experience and in line with growing concerns over water pollution, the Liaoning Environment Project was initiated in the mid-1990s at a total cost of US$338 million (including US$110 million provided by the World Bank and the remainder generated from provincial and municipal governments and local enterprises). The project was completed in 2004. A Staff Appraisal Report (or project feasibility study) was first carried out jointly by the Bank and LUCRPO and a proposal was then sent to the Municipal Planning Commission for approval. At the national level, negotiations took place between the NDRC, the Ministry of Finance and the Bank. As with all Bank projects, negotiations were time consuming. Three years transpired from the beginning of negotiations in 1992 to the final disbursement of the loan in 1995.

With a permanent staff of 25 (in addition to 10 temporary staff), LUCRPO was responsible for project implementation, with municipal project offices in Anshan, Benxi, Dalian, Fushan and Jinzhou. In addition to day-to-day management responsibilities, LUCRPO also played a key coordinating role between municipal and central level agencies and sub-project offices. The key project objectives included

1   To protect the main water resources in Liaoning province including the Hun-Taizi River Basin.
2   To strengthen pricing policies and institutional arrangements for environmental protection, water pollution control, wastewater and municipal solid waste management.
3   To initiate measures for air pollution and cultural heritage asset management.[84]

The project had a number of sub-components located in different cities: water supply and pollution control in Jinzhou, wastewater treatment in Anshan and Fushun, air pollution in Benxi and solid waste management in Dalian. Heritage conservation formed an additional sub-component that aimed to improve the classification and monitoring of cultural relics such as the Liaoning Great Wall at Jiumenkou and archaeological sites at Niuheliang and Jieshigang (see Map 5.1).[85]

The policy and institutional initiatives included establishing appropriate pricing policies, creating utility companies to enhance the performance of environmental services and establishing an environmental fund to support industrial pollution control.[86] These initiatives underpinned the financial viability of the physical infrastructure components and also had important implications for local capacity building. Pricing reform was a condition attached to the actual disbursement of funds in 1995. But despite small tariff adjustments in 1994, water and wastewater charges did not increase in Liaoning until 1998, and only then reportedly because the Liao (Liao River) had been singled out by the State Council in 1997 for a major clean up by the year 2000.[87] With limited funds, the provincial government had little choice but to impose user fees.[88] To overcome any local resistance to paying for clean-up costs, households and industry were billed jointly for water supply and wastewater treatment. In 1999, water tariffs in Shenyang stood at RMB1.5 per cubic metre (m^3) (including RMB0.5 for wastewater treatment) – an increase of RMB0.40 over 1994 tariffs.[89]

According to officials at the Bank's project implementation office, tariff reform had been delayed because, 'the Chinese government [was] concerned

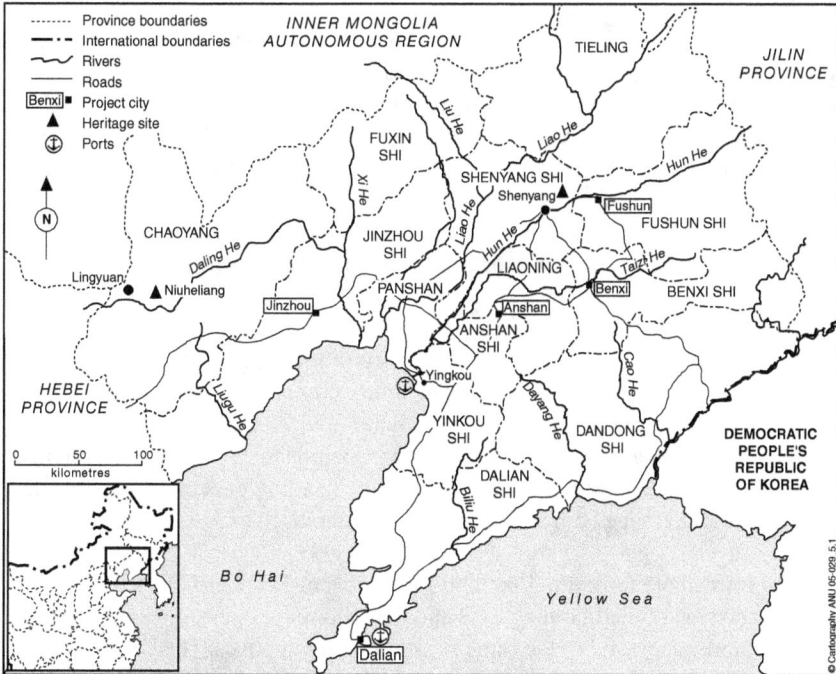

Map 5.1 World Bank Liaoning environment project 1995.

Source: Adapted from World Bank Staff Appraisal Report No. 12708-CHA, *China Liaoning Environment Project*, Washington, DC: Urban Development Sector Unit, East Asia and Pacific Regional Office, July 1994.

about inflation and imposing too much of a cost burden upon users too soon'.[90] Consequently, the tariff increases did not reflect the costs of operating and maintaining environmental services. In Shenyang, the real cost of the collection and treatment of wastewater is around RMB1.5 m^3 – three times in excess of the increased user charge. However, at the time of my visit to the project site it was clear that the Bank's pricing credo was becoming increasingly popular as local government officials began to grasp its economic significance. The remaining concern was that the system of joint billing did not allow the user to appreciate the costs of wastewater treatment and, therefore, the tariff was not acting as a disincentive to reduce discharges – this was particularly true for industrial enterprises, which given the right incentive could introduce measures for wastewater recycling.[91]

On a more negative note, the incremental approach to tariff reform was having serious implications for the six new utility companies created for wastewater and solid waste treatment. Clearly, without the necessary increase in user charges, these companies located in Anshan, Fushun and Dalian (each with a staff of around 200–500 people) were not commercially viable. Moreover, despite repeated attempts by the Bank to promote financial autonomy, the companies remained under the tutelage of the local government. A public water supply company was collecting fees and the portion for wastewater treatment was then handed over to the Municipal Finance Bureau. The latter repaid almost all of the funds to the drainage companies, together with a subsidy to cover the shortfall in operating costs. Local officials at the sub-project offices in Dalian and Jinzhou expressed serious concerns over the long-term viability of the companies, which they perceived to be inherently weak, or, to use a Chinese expression, *mianbao* (bread with nothing in between).[92] But officials in Shenyang were more optimistic. In the words of the deputy-director of the operations division at LUCRPO:

> Although we have experienced some problems in setting up the utilities and they are still relatively weak, it is important to recognize that progress has been made in corporate thinking. The utility companies now understand very well the importance of sound financial and institutional management. Moreover, they are now in a much stronger position to inform local government of their needs. Instead of the World Bank persuading government about what to do, the drainage companies can do it for themselves.[93]

In the immediate term, the Liaoning Environment Fund, established under the joint supervision of the Liaoning EPB and the Liaoning Finance Bureau, was seen to be the more successful of the Bank's institutional initiatives. The US$10 million fund provided loans to industries to implement cleaner production measures. The Bank provided an initial endowment of US$4 million with local cost-sharing funds drawn from pollution fees collected from industry. The fund was 'revolving' in that it required loans to be repaid in three years with a two-year grace period. With interest rates approximately 0.5 per cent below the rates offered by Chinese Banks, the fund was proving to be relatively successful with some enterprises showing a 10–20 per cent return on investment.

According to project officials, a key determinant of success was the extent to which market considerations were taken into account when allocating loans for cleaner production purposes. The state-owned Jincheng General Paper Mill in Jinzhou provides a good example of how the loan should work in practice but it remains an exceptional case. The paper mill discharges red liquor that is highly toxic. It cannot afford a wastewater treatment plant. But with the Bank's assistance, the red liquor is now being recycled to produce binders that can be sold to a number of factories in Guangzhou, Jilin and Tianjin. The mill, therefore, has been able to reap a return on its investment. This would not have happened in the absence of a detailed market analysis carried out by the enterprise management in conjunction with the World Bank.[94]

A potential problem with the fund is that enterprises are often slow to repay the loans, which are guaranteed both by the provincial and municipal governments. According to one EPB official, as far as the provincial government is concerned, 'money from one pot or the other is all the same thing, at least they are investing in environmental protection'. The issue of whether the enterprises are taking their loan obligations seriously is not considered to be important.[95] This attitude will hardly suffice in the longer term. At some stage the loans will have to be repaid in order for the fund to develop.

In regard to the implementation of the project as a whole, I found that local perceptions were largely favourable. Officials maintained that it had enhanced local understanding of the need for a macro approach towards environmental management. In the words of the deputy-director at LUCRPO:

> The Chinese approach towards urban environmental management is highly compartmentalized, with no overview of the system as a whole. The Bank takes more of a holistic approach towards understanding overall management needs by identifying the key issues on the basis of economic studies. Learning this systemic approach has been very important.[96]

In addition, they claimed that environmental services in Liaoning had significantly improved, thus creating a more conducive environment for infrastructure investment. However, timing was considered to be of the utmost importance. The Bank's Liaoning project demonstrated the difficulties involved in investing in physical infrastructure and developing the necessary policy and institutional arrangements at the same time. Ideally, the latter should come first. But the scale and severity of environmental problems in the immediate term is a constant pressure against adopting a more incremental and long-term strategic approach towards environmental management. This dilemma also raises an important capacity building issue. If institutional reforms are imposed too quickly there is a risk that local officials will not be able to grasp their significance. Without local understanding, institutional reforms and their economic objectives will undoubtedly fail.

Clean-up of the Dianchi Lake

Yunnan province ('South of the Clouds') is situated in China's southwest, bordering Vietnam, Myanmar and Laos. The province is 96 per cent mountainous and abundant in water resources with 40 plateau lakes and 6 river systems. Rich in biodiversity, Yunnan is relatively poor economically. The growth of industrial enterprises coupled with enduring poverty and population pressures have led to serious environmental degradation ranging from deforestation, soil erosion and loss of biodiversity to garbage build-up and water pollution. By 1997, 40 per cent of Yunnan's lakes and river systems were seriously polluted – water quality had declined below the lowest national standard (grade V).[97]

In 1996, Yunnan's water pollution problem received attention in China's ninth Five-Year Plan that, in turn, stimulated the local implementation of the 1,369 plan (100 factories to be treated, wastewater capacity to increase by 300,000 tons per day, 6 rivers to be cleaned up and 9 plateau lakes to be protected – all by the year 2000).[98] Notwithstanding the fact that these goals were overly ambitious for such a short time period, Yunnan's capacity for environmental management falls far short of its planning targets. This is hardly surprising when one considers that until recently the mandate of the provincial EPB was simply to deal with water supply and sanitation issues whereas now it has to develop and implement a province-wide environmentally sustainable development strategy. The Yunnan EPB has a similar number of staff (70 in total) as its counterpart agency in Liaoning (90 in total) but fewer resources and more complex environmental issues to deal with involving a wider range of stakeholders. Consequently, environment officials are convinced that 'strategic planning is the most important challenge' that they now face, together with the need for concrete action plans and public environmental awareness campaigns.[99]

Local government attention and resources to date have largely focused upon the Dianchi Lake, which became a state priority environmental project in 1996. The Dianchi lies downstream from Kunming. It is the province's largest watershed, with a catchment area of 2.4 million people. The lake is multi-purpose in providing 50 per cent of Kunming's potable water supply, and much-needed water resources for industry and agriculture. In the early 1970s, the lake was described by China's former Premier, Zhou Enlai, as 'the bright pearl of the Yun-Gui Plateau' and is still nicknamed 'the cradle' by locals. But the lake has changed dramatically over the past couple of decades or more. In 1975, the Dianchi harvested approximately 10,000 tons of fish per annum; rural residents used the lake for cooking and washing, farmers for irrigation and city residents for recreation. By 1995, fish stocks had been depleted by 60 per cent and the water quality of the lake had deteriorated to such an extent that it posed a serious threat to the health of the local population.[100] The rapid decline in water quality has been caused by a build-up of nutrients including phosphorous, nitrates and toxic heavy metals. High phosphorous levels have stimulated the growth of algae leading to a loss of dissolved oxygen and in turn eutrophication.[101]

Cleaning up the Dianchi does not pose any inter-jurisdictional problems, as in the case of many other watersheds in China, because it is under the single administration of Kunming municipality. The problem is that the sources of pollution are diverse, including municipal sewage and wastewater, industrial discharges, fertilizer run-offs and pesticides, and village wastes and are, therefore, difficult to manage. Before the 1990s, Kunming had no sewage treatment plants; wastewater (both residential and industrial) flowed untreated into the lake. Since then treatment capacity has increased with the construction of four sewerage plants financed by local and foreign funds but it is still insufficient to meet demand. Moreover, many industries still do not have their own water treatment or recycling facilities. TVEs surrounding the lake are a particular problem because they are very difficult to control. According to the Yunnan EPB, the TVEs refuse to comply with any government regulations in the belief that they are private and, therefore, have 'nothing to do with government'. Consequently, the Yunnan EPB has not been able to collect pollution fees from these enterprises.[102]

Agricultural or non-point sources of pollution are as difficult to control as the TVEs. Farmers traditionally travelled to Kunming and brought back nightsoil from pit latrines to use as fertilizer but now they are rich enough to use chemical fertilizers and have no intention of returning to the old ways. Furthermore, the use of pesticides has dramatically increased as farmers have shifted into more labour-intensive cash crops such as fruit and flowers. Village residents living along the lakeshore create an additional problem in that they no longer recycle their organic waste but instead dump it into the lake together with their accumulated plastic rubbish.[103]

From an historical perspective, land reclamation has also played an important role in the decline of the lake's water quality. Political campaigns during the 1950s to expand cultivated land by filling in surrounding wetlands led to a loss in the lake's nutrient absorptive capacity. Since the 1970s the water surface of the inner Caohai Lake has been reduced by one-third (from 12 to 8 km^2 in the 1990s).[104] Land reclamation (now illegal) still continues. Residents reclaim land and sell it to business people looking to set up restaurants and hotels by the lake. By 1998, over 70 hotels had been established, all dumping polluted water, sewage and rubbish directly into the lake.[105]

Since the late 1980s the local government has introduced a number of measures to clean up the lake, ranging from water pollution regulations to a government ban on the use of phosphorus detergents, and the construction of a dyke between the inner and outer areas of the lake.[106] But the central challenge remains one of facilitating cooperation between a diverse range of local polluters with limited environmental sensibilities. In particular, farmers and village residents are unwilling to take any responsibility; they believe that pollution is a result of urban and not rural development. In the case of the Dianchi, this distinction is blurred.

The World Bank Yunnan environment project

World Bank discussions over environmental issues in Yunnan began in the early 1990s. In 1995, at the request of the Bank, the British Overseas Development Administration (now the Department for International Development) funded a two-year programme to prepare Yunnan's loan application to the Bank. Project preparations focused upon the development of a Dianchi Environment Action Plan (DEAP). The plan recommended that the reduction of phosphorous concentrations in the lake should be a top priority for the Bank with the implication that improvements were needed in municipal wastewater management and industrial pollution control.[107]

The loan agreement was signed in 1996 at a total cost of US$349 million (including US$125 million from IBRD, US$25 million from the IDA, with the remainder coming from provincial and municipal governments and local enterprises). The project became effective in 1997 and is due for completion by 2006. The Yunnan provincial government appointed a leading group to oversee the project headed by the governor of the province. Project management is the responsibility of the Yunnan Project Office situated in the provincial EPB. Sub-project offices were also set up in Kunming, Gejiu and Qujing. The project aims to improve water quality and strengthen urban environmental service management in Yunnan province. Key project objectives are:

1   To strengthen policies and institutional arrangements for pollution control and municipal environmental services.
2   To support the improvement of water quality in the Dianchi Lake.
3   To facilitate sustainable investment in urban environmental services.
4   To introduce a comprehensive approach to urban infrastructure investment.[108]

The specific project components are divided into four key areas: Dianchi water quality recovery (involving wastewater treatment plants in Kunming and the county towns of Chenggong and Jinning), industrial pollution control (involving wastewater treatment at the Kun Yang Fertilizer Plant and the Kunming Chemical Fertilizer Plant), urban environmental service management in Qujing and Gejiu municipalities and institutional development (see Map 5.2).

As in the case of the Bank's project in Liaoning, policy reform and institutional development are a central focus of the project. Drawing upon its previous experience in Liaoning and Shanghai, the Bank supervised the creation of utility companies and tariff reform prior to the disbursement of the loan. The Kunming Drainage Company was established in 1995, with 500 staff, for the collection and treatment of municipal sewage. Two water supply companies were established in Gejiu and Qujing (each with 200 staff). But by 1999, the utility companies were still not financially autonomous; all three companies were continuing to remit revenues to their respective Municipal Public Utility Bureau.[109]

Tariff reform has been more successful, although tariffs still do not reflect the real costs of environmental services. Mindful of the low level of economic development in Yunnan, the Bank paid careful attention to the issue of affordability.[110] In 1995, at the request of the Bank, water tariffs (both industrial and residential) doubled from RMB0.32 m³ to RMB0.60 m³ (accounting for 0.6 per cent of local per capita income). Wastewater tariffs of RMB0.18 m³ were included in the total cost. Further tariff adjustments were made in 1998, bringing the total cost of water to RMB0.95 m³ (including a RMB0.35 m³ wastewater charge).[111]

But despite the satisfactory local response to the Bank's overtures in relation to pricing reform, and notwithstanding the fact that the Bank succeeded in implementing price incentives in a poorer region of China, these charges were not enough. In Yunnan, water and wastewater charges are only half of the pricing equation. The other half of the equation is related to agricultural prices. The fertilizer industry is one of the few remaining industries in China with strict price controls as a consequence of the central government's policy to hold down agricultural input prices. The industry is heavily subsidized but the subsidies do not cover operation and maintenance costs. Consequently, in the Dianchi catchment area, not only are the fertilizer factories operating at a loss, which does not bode well for future environmental investment, but they are doing so on the basis of perverse pricing incentives. It is, therefore, difficult to see how the use of fertilizers can be reduced.[112]

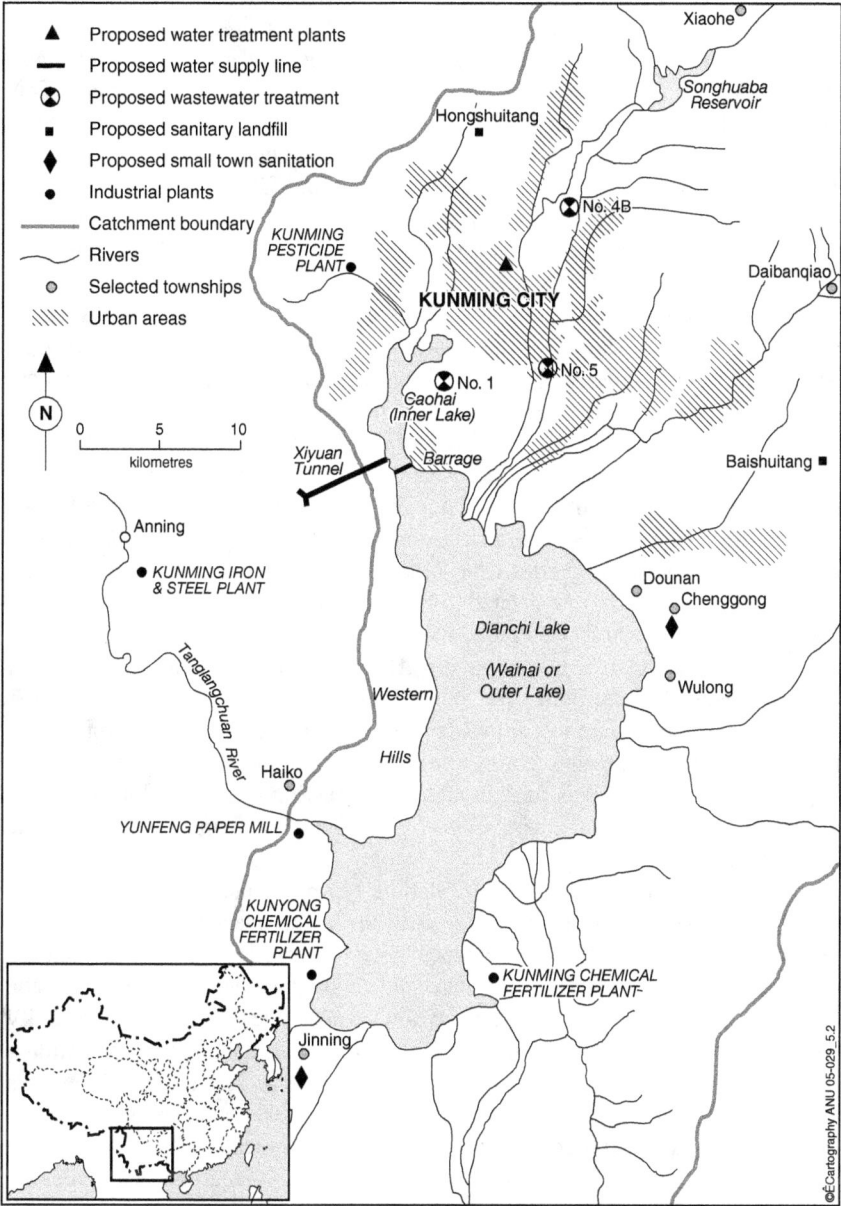

Legend:
▲ Proposed water treatment plants
— Proposed water supply line
⊗ Proposed wastewater treatment
■ Proposed sanitary landfill
♦ Proposed small town sanitation
● Industrial plants
— Catchment boundary
〜 Rivers
○ Selected townships
⫽⫽⫽ Urban areas

N

0   5   10
kilometres

Xiaohe

Songhuaba
Reservoir

Hongshuitang ■

⊗ No. 4B

KUNMING
PESTICIDE
PLANT ●

Daibanqiao

KUNMING CITY

▲

⊗ No. 1
Caohai
(Inner Lake)

⊗ No. 5

Baishuitang ■

Xiyuan
Tunnel

Barrage

Anning ○

● KUNMING IRON
  & STEEL PLANT

Dounan ○
Chenggong ○

♦

Dianchi Lake

(Waihai or
Outer Lake)

Wulong ○

Western

Hills

Haiko ○

Tanglangchuan River

YUNFENG PAPER MILL ●

KUNYONG
CHEMICAL
FERTILIZER
PLANT
●

● KUNMING CHEMICAL
  FERTILIZER PLANT

Jinning ○

♦

©ECartography ANU 05-029_5.2

Map 5.2 World Bank Yunnan environment project 1996.

Source: Adapted from World Bank Staff Appraisal Report No. 15361-CHA, *China Yunnan Environment Project*, Washington, DC: Urban Development Sector Unit, East Asia and Pacific Regional Office, May 1996.

During my visit to the project site, local project officials insisted that despite the poor economic performance of the lake's two major fertilizer factories, the Kun Yang Fertilizer Plant and the Kunming Chemical Fertilizer Plant, these enterprises were taking their environmental responsibility very seriously. This changing mindset had less to do with the Bank's intervention and more to do with the fact that between 1 May and 31 October 1999 Kunming hosted the World Horticultural Show. At a total cost of RMB10 billion (US$1.2 billion), the event was seen by the Chinese government as an opportunity to promote environmental awareness, in addition to stimulating international environmental exchange – a total of 67 countries participated.

For the purposes of this discussion, this event was important because prior to the opening on 1 May 1999, the State Council had issued a countdown plan (*lingdian jihua*) that ordered 253 key polluting enterprises in the Dianchi catchment area to comply with wastewater discharge standards by 30 April 1999 or otherwise be closed down. Three industries were in fact closed down, one paper mill was relocated, and 249 (mostly TVEs) complied with the directive. The crackdown had a positive effect upon the two major fertilizer plants involved in the Bank's project. The future of these enterprises was no longer certain and consequently, in the eyes of the beneficiaries, the World Bank environmental loans became equivalent to 'life loans' that needed to be taken very seriously.[113]

The central government's crackdown on polluting industries in Kunming also affected the implementation of the Bank's US$10 million environmental fund, co-financed by the Bank, the Yunnan provincial government, and a number of enterprises (which had to contribute at least 30 per cent). The reason that so many enterprises managed to comply with the central directive was because they had been able to access this fund in addition to relying upon the Chinese banks. It remains under the joint supervision of the Provincial Finance Bureau and the provincial EPB.

The World Horticultural Show also had its environmental downside. Damage limitation measures were taken in the lead up to the event that wasted much-needed resources. Both the central and local governments provided RMB250 million to remove algae blooms from the lake and to dredge sedimentation from the Caohai. This was little different from the remedial action taken to beautify Kunming City, only recently recovering from a huge construction binge, by building flower beds with thousands of flower pots that were replaced every three days for six months! Throughout 1999, the 'flower pot phenomenon' in Kunming undermined the development of any long-term environmental action.

Any negative impact on the Bank's environmental project, however, appeared to be minimal. According to EPB officials, in comparison with foreign-funded environmental projects in the past, the Bank's project was relatively successful. In the early 1990s, for example, the Australian Agency for International Development (AusAID) had funded a sewage treatment plant without a drainage network, which meant that the plant had no carrying capacity and, therefore, could not be put into operation.[114] At a minimum, the physical works involved in the Bank's environmental project were reportedly effective. Moreover, project officials maintained that their understanding of the financial aspects of environmental management had greatly increased. In principle, the need for pricing reform struck a chord with local officials but, as stressed by project officials in Liaoning, timing was considered to be the crucial issue. In the words of one project official:

> It is difficult because we have to take social issues into account. Also China has a long history of government planned prices, we cannot change overnight. Basically, central *and* local governments agree with the World Bank stance – market reform is part of the national mandate – but it must be gradual reform.[115]

But local EPB officials had one additional reservation that put the entire project into question: they believed that the project should have been more centred upon solving non-point source control and cleaner production. In the words of the deputy-director of the project office:

> The World Bank prefers to do point source control because it is easier to measure. The perception of this office is that the task managers involved in the project are from the urban environmental division of the Bank and, therefore, they are not interested in rural issues. It is also true that analysis on rural issues is difficult to define and carry out.[116]

At the request of the provincial government, the Bank agreed to fund a pilot rural sanitation project. At a cost of US$2.3 million, the project involved three villages – Xiaohe, Dounan and Wulong (see Map 5.2). The project aimed to 'disseminate affordable, village-level approaches to waste management' involving wastewater control, improved agricultural waste management practices, nightsoil utilization, and solid waste control. But the major issue of pesticide and fertilizer control was not addressed for two reasons. First, institutional capacity was weak. The Soil and Fertilizer Station of the Kunming Agricultural Bureau was responsible for implementing the project, but it had no experience in pesticide control and eco-agricultural practices.[117] Second, the implementing agency had no authoritative power to address the issue in the face of strong resistance from village residents. On paper the villages are expected to pay up to 30 per cent of the investment costs but village residents have outwardly refused to accept any responsibility, financial or otherwise, in the belief that the clean-up of the Dianchi is primarily the responsibility of industry and the municipality of Kunming.

The fundamental problem here is the blurred distinction between rural and urban environmental management in the Dianchi catchment area. The villages are essentially semi-urban settlements with relatively large populations (between 10,000 and 30,000 residents) and they have close economic links with Kunming City.[118] The settlements look like small townships but they lack the necessary physical and managerial infrastructure to support urban economic activities and to safeguard the environment. Although reliable figures are not available, the total amount of organic and inorganic wastes (sewage, wastewater, chemical fertilizers and pesticides) generated by the settlements is deceptively high. As a consequence of China's restrictive policies on urban registration, the residents do not have urban status which, in turn, increases their unwillingness to share responsibility for the clean-up of the Dianchi with registered urban households in Kunming.

Under such circumstances, the Bank's pricing approach is clearly not sufficient to improve environmental management because it does not address the need to enhance shared responsibility. To this end, a participatory approach is more likely to work, particularly if it is linked to the issue of local ownership. In this regard Yunnan is a step ahead of other provinces in China in that the groundwork has already been laid. Approximately 15 Chinese research institutes are involved in participatory development in Yunnan. The approach is widely used in rural areas with notable success. A good example is one Miao village that is organizing its own trust fund for developing water supply and roads. The villagers themselves have agreed upon the prices for water and road tolls. The cost of water (potable water as opposed to water for irrigation purposes) is higher in this village than the price for water recommended by the World Bank in Kunming.[119] In the case of the World Bank project, not only is a participatory approach needed to convince villagers of the need to pay for resources, but local ownership is also needed to ensure the sustainability of environmental investment. The rural sanitation component is likely to fail on two accounts: first, because the beneficiaries are unwilling to share investment costs; and second, because not enough attention has been paid to the issue of transferring the assets to the villages on completion of the project.

Overall, the Bank's experience in Yunnan challenges the clear-cut distinction between urban and rural environmental management that is characteristic of the Bank's environmental lending.[120] The Bank seemed impervious to the local realities of managing environmental problems in poor semi-urban settings. It completely failed to grasp the strategic importance of dealing with non-point source pollution – an issue that was perceived to be vitally important by the local government. Consequently, it was not clear that the Bank's environmental assistance would have any substantial impact in the longer term. As acknowledged in the small print of the Bank's Staff Appraisal Report, there was a relatively high risk that 'non-point source pollutants from agriculture, TVEs and villages could overwhelm any benefit that would otherwise be realised by the [Bank's] planned investments'.[121]

Beyond traditional practice in Guangxi Zhuang Autonomous Region

In light of the preceding discussion, the Bank's environmental project in Guangxi is particularly interesting because it does recognize the limitations of a market-oriented approach to environmental management in poorer and less concentrated urban settings. This realization largely came about as a result of the Bank's previous experience of working in the poorer areas of other cities in China.[122] In this project, the Bank's primary focus upon pricing and market-driven institutions is broadened to include participatory practices. In contrast to the UNDP's participatory projects, the focus is upon decentralized decision-making at the community level rather than interagency coordination. The approach, however, is still aligned to the Bank's central concern with cost recovery. The assumption is that involving beneficiaries in the design, implementation and financial management of environmental facilities will ensure greater sustainability (both economic and environmental) in the longer term.

Environmental dialogue between the Guangxi regional government and the World Bank began in 1994. A project agreement was signed in April 1998, at a total cost of US$175 million ($72 million IBRD and $20 million IDA), and the project is due for completion at the end of 2005. A significant part of the project focuses upon providing environmental services to the urban poor.[123] This was inspired by the Bank's earlier involvement in a poverty alleviation project in southwest China in 1995. The latter raised awareness of the large migrant population in Guangxi (estimated to total 20–30 per cent of the urban population) that were living in old neighbourhoods in Nanning and Guilin, as well as in former rural villages in the urban peripheries, with extremely poor environmental services.

The central focus of the project, however, is to improve the quality of the region's main water bodies – the Lijiang (Li River) and its tributaries in Guilin, and the Chaoyang stream and Yongjiang (Yong River) in Nanning (see Map 5.3). The quality of the water in the Lijiang – an 83-km waterway that stretches between Guilin city and Yangshuo county – has particular significance because the river is a major tourist attraction and, therefore, a vital source of local government revenue. The Lijiang suffers from low water levels during the dry season together with relatively high pollution loads. A major aim of the project, therefore, is to increase the flow of water in the river in order to enhance its carrying capacity[124] while, at the same time, reducing the inflow of raw municipal sewage and wastewater. Key project objectives include:

1   To upgrade sewage, drainage and solid waste systems in Nanning and Guilin cities.
2   To improve the region's institutional and financial capacity for environmental protection, environmental services and water resource management.
3   To pilot a participatory approach to environmental improvement in poor neighbourhoods.
4   To implement pilot schemes for controlling pollution from sugar refineries.[125]

The Guangxi Urban Environment Project Office (GUEPO), situated in the regional EPB, is responsible for implementing the project. GUEPO is under the supervision of the vice-governor of the region and headed by the vice-director of the regional EPB. Four other project offices have been established under the municipal EPBs in Nanning and Guilin and the county EPBs in Yangshuo and Lingui. Although at the time of my visit the project was at an earlier stage of

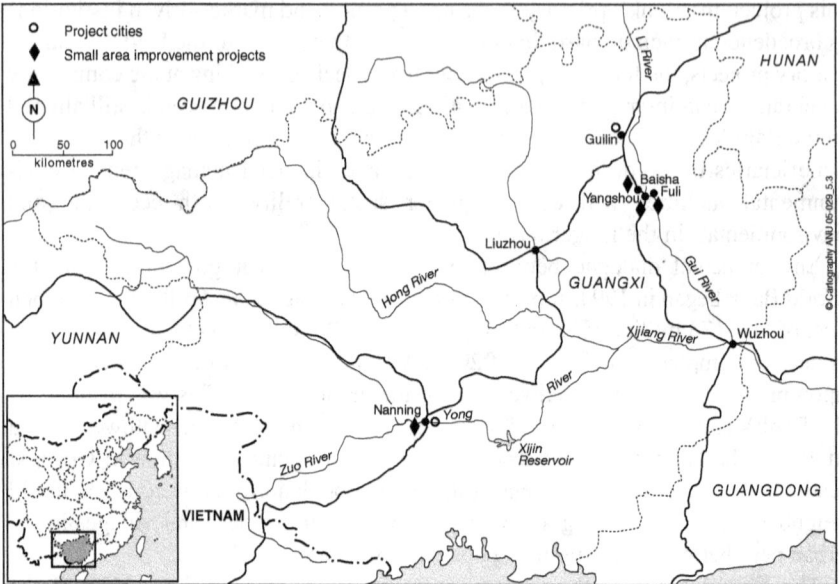

Map 5.3 World Bank Guangxi environment project 1998.

Source: Adapted from World Bank Staff Appraisal Report No. 16622-CHA, *China Guangxi Urban Environment Project*, Washington, DC: Urban Development Sector Unit, East Asia and Pacific Regional Office, May 1998.

implementation compared with the Bank's projects in Liaoning and Yunnan, some progress had already been made, especially in relation to financial management. As a condition of the loan disbursement, both Nanning and Guilin municipalities had introduced tariff increases prior to implementation. Residential water tariffs per cubic metre had increased from RMB0.32 in both cities in 1995 to RMB0.40 in Nanning and RMB0.42 in Guilin in 1998. Charges for wastewater had also been introduced for the first time at a cost of RMB0.12 m³ in Nanning and RMB0.18 m³ in Guilin – local residents were being billed jointly.[126] As in other cities in China, these charges were not sufficient to recover the full costs of operating and maintaining environmental services, but they were a step in the right direction.

In 1999, plans were also underway to establish two new sewage treatment companies in Nanning and Guilin. Both companies were hoping to start operations in 2000. Implementation is likely to be easier in Guilin than Nanning because the latter lacks any previous experience in sewage treatment, whereas the former already has a drainage construction division that is relatively effective in treating the city's wastewater and sewage.[127] Financial autonomy is likely to be problematic in both cities.

The bigger problem is that strengthening municipal environmental services is not sufficient to improve the water quality of the Lijiang because it stretches beyond concentrated urban areas. With many rural villages situated along the river, attention also needs to be given to pricing agricultural water use or to what the Bank refers to as 'raw water pricing'. To this end, the Lijiang Water Resources Commission has been established to coordinate water resource pricing and water distribution, in addition to coordinating the implementation of the Bank's investments in physical infrastructure. A vice-governor chairs the Commission with representatives from the relevant regional bureaux (the planning commission, the finance bureau, the EPB and the water resources bureau), and Guilin municipal government. It is, however, still too early to judge the effectiveness of the Commission in practice.

In regard to industrial pollution control, a sub-loan facility has been set up to provide financial support for reducing pollution discharges from sugar refineries that generate approximately 10 per cent of the region's industrial output and 75 per cent of its water pollution.[128] The US$5 million fund (including US$2 million from the Bank and US$3 million from the enterprises involved) provides loans that are repayable in seven years, including a two-year grace period, at a locally competitive interest rate. In contrast to the Bank's environmental loan facilities in Liaoning and Yunnan, the selection criteria are more stringent. Companies applying for the loan must be proven to be profitable, with clear financial projections for the coming five years, and a debt maintenance ratio of at least 25 per cent. A unit has been established under the regional EPB to screen applications with representatives from the Guangxi Sugar Company. However, local EPB officials have expressed concerns that the loan facility only appeals to a small minority because the sugar refineries in Guangxi are 'technologically risk adverse; they do not want to act as a technological pioneer'.[129] This lack of interest has something to do with the fact that these relatively small state-owned enterprises have been sheltered from any external competition – no new refinery has been built in Guangxi since the 1970s.

Despite this rather muted response from industry, after the first year of implementation the local government's response to the Bank's institutional and financial measures was generally positive. Like their counterparts in Liaoning and Yunnan, project officials in Guangxi were generally supportive of the Bank's market-oriented approach towards solving environmental problems in their region. Perhaps not surprisingly, they were far less supportive of the Bank's new focus upon community participation. From the very beginning, the Bank's move into the uncharted territory of community participation stimulated a puzzled response from the region's top officials. According to the Bank's task managers in Washington, local officials could not understand why environmental management needed to be taken out of the hands of local government. To promote participation amongst governing bodies, as in the case of the UNDP, was one thing, but to openly advocate community rather than government control over environmental investment was quite another.[130]

Traditionally in China, district governments are responsible for urban environmental services at the neighbourhood level and government-sponsored design institutes carry out the design and implementation of construction projects. The

Bank's intention was to turn the established system upside down by promoting community responsibility and the active involvement of the beneficiaries in project design and implementation. A small area improvement (SAI) component (at a total cost of US$9.2 million) involved upgrading environmental facilities, including sewers, toilets, garbage collection and access roads in poor neighbourhoods. To this end, the Bank proposed using a *demand-responsiveness* approach that had been tried and tested by the World Bank–UNDP Water and Sanitation Program.[131] Simply put, the *demand-responsiveness* approach advocates that rather than donor or government agencies determining local needs, the local beneficiaries themselves should express their demands through dialogue with government agencies and service providers. Actual demand can be measured by determining the users' willingness-to-pay for services either in monetary terms or in kind.

The implementation of this approach has proven to be time consuming with project preparations taking five years to complete. One inevitable consequence has been an increasing ennui over the issue of participation. 'To be frank', one project official observed, 'these projects are a big headache; the people are tired of the process and the project office is embarrassed at the slow progress.'[132] Lack of experience both on the side of the World Bank and the local implementation agencies in China, was a major factor behind the time lags involved. Sixteen SAI projects were eventually identified (8 in Nanning, 2 in Guilin and 6 in Yangshuo). Selection criteria were based upon local environmental conditions and the scope for future government investment. For example, Baisha (a former rural village outside Yangshuo County) had no drainage system, no sewage system and no landfill site; the township was overwhelmed by its own wastes. After many failed attempts to acquire funding from the county government, Baisha had few hopes left for securing healthy and environmentally sound living conditions.

In most cases, the participatory process has succeeded in increasing the local beneficiaries' willingness-to-pay. Table 5.1 provides an illustration of how in Bianyang neighbourhood in Nanning an original funding proposal evolved into a more realistic plan based upon the key issue of local affordability. When discussions first commenced in 1996 four priorities were selected (sewer, road, public toilet and garbage collection) at a total cost of RMB6.4 million. Residents were required to meet 25 per cent of the construction costs (RMB1.6 million) which they could not afford. Two years later, after an iterative process of planning, dialogue and redesign, priorities were narrowed down to a new sewer and an access road at a total cost of RMB2.15 million with a local community share of RMB0.56 million, representing 30 per cent of the costs. What is significant here is that the local government was no longer required to provide funding, thus placing a significant financial responsibility for the projects in local hands.

However, not all the projects involved followed suit. For all 16 SAI projects, the community share of investment costs ranged from 10 to 30 per cent. Moreover, actual implementation varied: the participatory process at the county and township levels was more community driven than at the municipal level. At

Table 5.1 Local beneficiary participation in Bianyang neighbourhood, Nanning 1996–1998

Project components and costs	07/1996	06/1997	11/1998
Project component	1 Sewer	1	1
	2 Access road	2	2
	3 Public toilet	—	
	4 Garbage collection	4	
Proposed funding			
World Bank	50%	50–70%	70%
Local government	25%	—	—
Community	25%	30–50%	30%
Community share	RMB1.6 million	RMB1.3–2 million	RMB0.65 million
Total project cost	RMB6.4 million	RMB4 million	RMB1.84 million

Source: Adapted from Guangxi Regional Project Office, 'Start-up workshop report for small area improvement component', prepared by the World Bank in consultation with the Regional Project Management Office for Guangxi Environment Project, Nanning, 4–6 February 1999.

the township level, priority listings generated from local consultations were sent to every household and then determined at a meeting between township leaders, neighbourhood committee representatives, women's groups and enterprise managers. The design process involved displaying site maps for sewers or landfill sites on the town walls and inviting verbal or written responses. A final meeting led to a consensus over the allocation of investment costs.[133] In contrast, in Nanning, government-designed surveys were sent out to households and priorities were determined on the basis of computer analysis. The priorities were then presented for final approval at a meeting between representatives from resident neighbourhood committees and the district and municipal governments. In the same manner, representatives approved the design work once it had been completed. A final meeting was held to determine the allocation of resources.[134] At both levels international consultants were used by the Bank to provide guidance.

Given this variance in implementation, it was not surprising that local responses to the approach differed widely between the project offices involved. In Nanning, project officials were overtly critical. Although they supported the concept, they thought that the method was 'too westernized'. One official critically observed:

> In China, the people are not used to arriving at solutions through argument; they are used to guidance from experts. You could ask many local people what should be done to improve environmental services and they would give

you a large number of different answers. But this is not practically useful. In China the government has carried out many small projects through consultation with local people, but experts make proposals and then the people select various options.[135]

A major criticism was the perceived mismatch between the approach and World Bank procedures in that the Bank had advocated community-driven participation but at every stage of the implementation process it had been 'too hands on' and 'interventionist'. Officials at the project office were obliged to seek approval from Washington over every aspect of the design and consultation process, but this involved long delays given that the Bank's task managers only visited the region twice a year. Moreover, it was argued that although the urban and environment unit in Washington was willing to be flexible over the Bank's rules and procedures, the financing division demanded the submission of annual output targets regardless of the process involved. The approach was also viewed as being unrealistic in that it underestimated the difficulties involved in shifting responsibility away from the government:

> The World Bank believes that government design units are too traditional in their approach. The first problem is that there is no legal framework or supervisory body in China to oversee the quality of independent design consultants. The central government has recently issued new regulations for quality control. Government design institutes are now responsible for a project for life. How responsible is an independent consultant? The second problem is that design institutes are paid set fees and cannot earn above a certain rate. A participatory approach is too time-consuming; it would mean that these units would be working for nothing![136]

Officials from the Guilin project office were more ambivalent – neither overtly critical nor supportive of the participatory approach. It was thought to be 'too complicated' but at the same time 'a useful experience that would allow for some improvement in environmental management in the future'.[137] But whereas the Bank's participatory agenda may have put some officials at the municipal level on the defensive, it struck a responsive chord amongst local officials at the county level. Project officials in Yangshuo were far more aware of the relevance and practicality of the approach. In the words of the director of the project office: 'the approach is useful because local residents can provide ideas to improve the design of projects. In addition, project implementation is smoother because of the local sense of ownership and responsibility.'[138]

In light of this variance in local perceptions, it would seem that the Bank's participatory approach is more applicable at the county and township level and precisely in those semi-urban settings that have been neglected by the local government. From a capacity building perspective, despite the difficulties involved, the approach is equally important for municipal settings because it encourages a local sense of environmental responsibility, which hitherto has not existed: traditionally the local people have relied exclusively upon the local government to solve their environmental problems. The issue of shared responsibility is not only important for the purposes of implementing environmental projects in China; it is vital to the success of China's efforts towards solving its environmental problems per se. This is what officials at the municipal level utterly failed to comprehend.

Institutional innovation, or *Potemkin House?*

Is the Bank's market-institutional approach to environmental management in China leading to institutional innovation or a *Potemkin House*? The answer can be interpreted in two different ways. If one simply matched the Bank's objective with the outcome on the ground one could conclude that the Bank's creation of market-oriented utility companies in China was little more than a façade: these entities are not financially independent and remain subordinate to the local government bureaucracy. But a different conclusion could be drawn if one were to ask the question: how is the Bank's environmental assistance contributing towards improving local capacity? The strength of the Bank's approach lies in building local financial capacity for environmental management by reinforcing the positive role that the market can play in solving environmental problems and by underlining the critical importance of economic viability. In this respect, an attitudinal change was taking place amongst local officials, especially in the more developed northeast of China, which signalled an important first step in the direction of institutional reform.

The project investigations discussed earlier suggest that the Bank's approach is more effective in the relatively developed northeast of China than the lesser developed southwest. This variance has more to do with spatial factors or *the degree of urban concentration* than purely economic factors. Despite huge income disparities, the Bank's policy of getting the prices right was embraced by local officials in both regions as an important means of establishing affordable environmental services. The problem for both poorer and relatively developed regions was not the concept per se but the issue of timing. In keeping with China's economic policy of 'gradualism', local officials preferred incremental price reform to sudden price hikes. Pricing reform was slow; but at least in comparison with earlier Bank projects it was in fact taking place.

Where the Bank came unstuck was over the issue of institutional development. The Bank's narrow conception of institutional development for environmental purposes (i.e. the creation and development of market-oriented environmental services) was not uniformly effective in different local settings. It was more appropriate in highly concentrated urban settings where environmental problems were confined within municipal jurisdictional boundaries and where institutional responsibility could be shared amongst traditional urban agencies. In northwest China, for example, the urban/rural divide is reasonably distinct with high concentrations of urbanization in the major industrialized cities. Consequently, environmental management involves cooperation between traditional urban construction, planning and environmental protection bureaux, together with largely state-owned industrial enterprises. In contrast, in southwest China, the urban/rural divide is more ambiguous with many previously rural villages evolving into semi-urban settlements on the outskirts of major cities. Consequently, environmental management not only involves traditional urban institutions, but also farmers, township and village enterprises, and village residents. The structural transformation that is bringing urban and rural areas closer together in China has clear economic benefits but deepens the complexity of environmental management by increasing the number of local stakeholders involved. Under these circumstances, incentives do need to go hand in hand with participation.

As the World Bank environmental project in Yunnan province clearly attests, getting the prices right and creating market utilities is not sufficient when environmental problems reach beyond traditional urban boundaries. In the latter case, private ownership is not the solution. Instead, a more strategic approach is needed to encourage greater participation at the grassroots and within the local bureaucracy. Institutional arrangements need to be created for managing resources more effectively and more attention needs to be given towards encouraging sectoral cooperation between the agricultural bureau, the planning bureau, the water resources bureau and the construction bureau. As it currently stands, the Bank's environmental project in Yunnan is having a negative effect upon capacity building, not only because it is ignoring local strategic concerns but also because it is encouraging local officials to divert much-needed resources into traditional point source control.

In both regions, the fundamental weakness of the Bank's market-oriented approach lies in its attempt to carry out institutional development and infrastructure investment simultaneously. Ideally, institutional capacity needs to be created before the construction of environmental infrastructure takes place to ensure that costs can be recovered. Without the institutional capacity to manage environmental services, both financial and environmental sustainability of such projects remain at risk.

Getting the prices right and creating market institutions is not the only contribution that the Bank can make towards facilitating environmental capacity in China. Despite the difficulties involved, the Bank's participatory approach has reaped some positive benefits. Given the innovative nature of the Bank's participatory agenda, it was plagued from the start by contradictions, among them the very notion of 'participation from above'. Local officials at the municipal level were fairly strident in their criticism of the Bank's tendency to 'micromanage' community participation in China. But here the Bank was coming up against the resilience of the Chinese old guard. The same criticism was not being made at the lower levels of government. More significantly, the Bank's task managers involved in the project were reportedly making a conscious effort not to intervene too closely. In the words of one task manager, 'the task teams go in and cause a stir but ultimately the people themselves must manage the project. A too strong supervisory approach is not going to work as it undermines confidence.'[139]

The more fundamental contradiction lay in the incongruity of a large financial institution such as the World Bank becoming involved in small-scale projects at the grassroots. In short, a long-term community approach was ill-suited to the Bank's established rules and procedures. The alleged tendency to 'micromanage' was, therefore, more a case of ensuring that financial obligations were being carried out rather than a deliberate attempt by the Bank to impose its will upon local beneficiaries. Despite this inherent contradiction, in less concentrated urban settings with more political flexibility, the Bank's participatory approach did succeed in promoting local ownership over environmental projects and, in turn, enhancing the local sense of shared environmental responsibility – an impressive outcome given the difficulties involved.

Conclusion

As in the case of Japan and the UNDP, the Bank's approach to environmental management in China is somewhat limited by the need to reconcile donor priorities with recipient needs and expectations. Indeed, the Bank has encountered the same difficulties as the UNDP in trying to promote participatory practices that involve Chinese citizens. The Bank has been able to introduce community participation to China on an experimental basis because this was nested within a larger project to build environmental infrastructure.

From the perspective of the Chinese government, the Bank's market solutions to environmental problems are more appealing. Increasingly, the Bank's market approach to environmental management is being taken seriously at both the national *and* local government levels. The uneven process of market reform in China together with the huge regional disparities in per capita income have not prevented poorer regions from implementing price reforms, although progress has been far slower than in the more developed regions of China. The Bank has had less success in implementing institutional reform in China because this requires a long-term commitment towards building institutional capacity and enhancing local awareness.

Overall, the World Bank's environmental assistance in China has amounted to more than a chimera but less than a complete change in institutional practice. But whether the Bank has promoted a visible change in institutional structures in China is not the only measure of success. The Bank has had a significant influence over local attitudes to environmental management on two levels. First, at the municipal level, it has convinced many local officials of the importance of creating financial incentives. Second, at the county and township level, albeit on a far lesser scale, it has cast the issue of responsibility in a stronger light by promoting a shared sense of financial responsibility for environmental improvement. These attitudinal changes are less tangible than the creation of new market entities but equally important, if not more so, for local capacity building over the long term. For the Bank's approach to succeed in the future, what is now needed is greater institutional flexibility, both on the part of the donor and the implementing agencies in China.

6   The promises and pitfalls of international environmental aid to China

The three case studies in this volume convey a graphic impression of both the promises and pitfalls of international environmental aid to China. On the one hand, despite difficult circumstances, international donors have had some success in strengthening local capacity, especially in relation to financial efficiency and interagency coordination. On the other hand, progress in building *sustainable* environmental capacity has been constrained by certain local conditions as well as weaknesses on the part of the donor institutions.

Since a major objective of this volume is to increase our understanding of the relationship between different international donor approaches to environmental management and local capacity, this chapter begins by highlighting some of the insights from the case studies that draw the relationship together. In so doing, the conceptual framework proposed in Chapter 1 is further refined and developed by stressing the importance of shared responsibility. This social dimension tends to be underestimated by scholars and practitioners alike but it plays a crucial part in the overall development of environmental capacity.

The analysis then shifts to comparing the local conditions under which each donor approach is effective. It is significant that the donors examined in this study had made little effort to adjust their approaches to diverse socio-economic circumstances. Yet, the empirical findings clearly demonstrate that the effectiveness of different donor approaches varies considerably across regions. Variance is not simply related to the level of economic development as one might expect, but also to political, institutional and spatial factors. Hence, a key determinant of effective international environmental aid seems to lie in matching different donor approaches with appropriate regional settings.

Finally, I will examine the broader limitations of environmental capacity building from the donor perspective. Common constraints involve a low level of institutional commitment, an overemphasis upon project preparation rather than implementation, poor coordination between environmental grants and loans, and a high level of institutional inflexibility. These constraints are further aggravated by the ubiquitous problem of poor inter-donor coordination.

The relationship between international donor approaches and local environmental capacity

Local responses to donor-funded environmental projects can be taken as a baro-meter that registers the level of commitment towards the environment. Although weak compliance is still a problem at the local level in China, resistance is by no means uniform. In different regions of China the outlook is more varied. By and large, local officials are aware of the economic and social costs of ignoring envi-ronmental problems, but they are often overwhelmed by the difficulties involved in striking a balance between economic development and environmental protection. In many cases, the political will does exist but the capacity – institutional, financial and technological – does not.

International environmental aid is a barometer in other ways as well, in registering the degree of local government support for both market and human development approaches to solving environmental problems. State-centred environmental protection is still dominant in China but this is by no means a static paradigm. Findings in this study reveal a shift at the central *and* local levels towards a more comprehensive notion of environmental management that embraces the importance of the market in particular. Although a more democratic style of participation that involves citizens in the decision-making and imple-mentation of environmental policies and practices remains a distant goal, the notion of stakeholder participation at the elite level is beginning to take root. What is important for the purposes of this study is that environmental donors have played a significant role in facilitating this shift. This, in turn, has had important implications for capacity building. Local officials are now less fixated upon engineering solutions for managing environmental problems and more aware of the need to build institutional and managerial capacities.

Developing the conceptual framework

Although local environmental capacity in China varies considerably across regions, many of the weaknesses identified in this study are little different from the capacity constraints found in other developing (and sometimes developed) nations – namely inefficient financial management, poor interagency coordina-tion, backward technological knowledge and a weak environmental information base. Hence, the instrumental aspects of environmental capacity (financial efficiency, institutional cohesion, technological advance and information sharing) that were identified in Chapter 1, are no less relevant to China. This is in keeping with the expectations of scholars working on the issue of environmental capacity in Eastern and Central Europe.[1]

But at a deeper level, this study has demonstrated that if the purpose is to build environmental capacity that can endure over time then we need to pay more atten-tion to its social dimension. Recall from the theoretical discussion in Chapter 1 that the informational aspect of environmental capacity was considered to be analytically weak without an understanding of how information sharing actually changes behaviour in any given social context. In taking into account China's particular political and socio-economic circumstances, the discussion that follows will highlight the relevance of the instrumental dimensions of environmental capacity as well as consider how and under what guise the social dimension can be fully incorporated into the overall framework.

In China, as in other developing countries, a lack of funding for environmental management is the most obvious constraint upon environmental capacity. This involves the capacity for resource mobilization and resource utilization. Not surprisingly, in the poorer regions of China access to funds is deemed more important than their efficient allocation. However, the problem of cost-recovery in relation to environmental services and investment is now more widely recognized. Consequently, the World Bank's emphasis upon appropriate pricing structures has been well received in both poor and relatively developed regions. Utility prices remain below the real costs of operating and maintaining environmental infrastructure and services, but considerable improvements have been made which bode well for future environmental investment. Furthermore, under the guidance of the Bank, the establishment of environmental funds has boosted enterprise investment in pollution control. Local governments are also making efforts to screen applications for environmental loans and conduct feasibility studies in order to ensure their effective utilization. The problem is that old habits die hard. Despite a growing recognition of the need for market competition in China, many SOEs still view an environmental loan as a gift rather than a financial obligation. The situation is made worse by the fact that government agencies controlling the funds have no incentive to ensure that environmental loans are repaid. This means that the efficient utilization of financial resources in China is unlikely to take place without increased market competition and financial accountability.

As in the former communist countries of Eastern and Central Europe, building institutional capacity for environmental management in China is particularly important because of the continuing legacy of central planning and the lack of experience in creating partnerships amongst different stakeholders. As revealed by the UNDP's environmental projects in China, the capacity for interagency coordination is largely dependent upon the seniority and political status of the actors involved. But despite political constraints, comprehensive environmental planning – both within and between government agencies – is improving in response to growing economic and social pressures. As the economic and social costs of ignoring environmental protection increase, local governments are forced to take preventive action. Moreover, public pressures on governments, at all administrative levels, are now widespread.

China's institutional capacity for openness, however, remains weak. In this respect international donors have been less than successful. Some improvements have been made in respect to public consultation, especially in the cities of Shenyang, Benxi and Guilin, but the exchange of ideas and the flow of information between government agencies is still seriously hampered by traditional administrative structures that are based upon narrowly defined sectoral divisions.

China's technological capacity is less constrained by country-specific circumstances. The capacity for knowledge creation and acquisition relies upon improved monitoring procedures, information management and environmental auditing expertise, which are relatively easy to improve over time. Developing the capacity for technological diffusion is less straightforward. First and foremost, as in any other developing country, it relies upon an enhanced awareness of both the economic and environmental benefits to be gained from financial investment. New technologies must be environmentally sound *and* cost-effective. With the support of international donors (UNDP in particular), China has made some progress in disseminating the benefits to be gained from technological acquisition. However, the potential for technological diffusion lies more in a long-term strategy that relies upon 'learning by doing' and the actual demonstration of the economic benefits to be gained from investing in environmental technologies.

Information sharing is an important element of developing environmental capacity in China, but it is less important than the broader social dimension. In the literature on environmental management, the emphasis is upon a linear process of information sharing or disclosure strategies leading to greater partici-pation and, in turn, attitudinal change.[2] In this study, while information exchange was clearly important, especially in the city of Wuhan, changing attitudes on the part of local officials were largely brought about through a socially interactive process of building environmental constituencies and nurturing environmental awareness through active participation. As argued in Chapter 1, information disclosure does not necessarily guarantee a change in behaviour. To this end, we need to pay more attention to social norms. This study throws light on the social dimension of environmental capacity by stressing the importance of *shared responsibility*.

In China, political commitment towards protecting the environment at all levels of the government administration is increasing: environmental laws and regula-tions are expanding and financial investment is high relative to other developing countries. The downside is that there now exists an implicit understanding amongst city residents, enterprises and farmers alike that the government is solely responsible for protecting the environment. Despite government efforts to build environmental awareness through public consultation, educational activities, and even the Internet, the sense of public responsibility for the environment remains low. Part of the problem is that environmental NGOs in China are still heavily controlled by the government, and citizen involvement in environmental manage-ment remains weak. But it is also the case that for many Chinese the environment is viewed as a public health issue and it is, therefore, little different from any other public service to be provided and maintained by government agencies. The idea that the responsibility for the environment must be borne by the populace as a whole has yet to permeate beyond the classroom and elite intellectual circles.

Developing a sense of shared responsibility is equally important in relation to local government agencies, enterprises and farmers. Although the government in China has the main responsibility for protecting the environment, this usually falls heavily upon the rather weak shoulders of the EPBs. An important barrier to effective environmental management in China is reluctance on the part of other government agencies to accept responsibility for environmental planning and share budgetary resources accordingly. A second barrier is recalcitrance on the part of industrial enterprises. Environmental regulations (when they can be enforced) together with point source control may be effective in producing quick results, but far more is needed to encourage enterprises to accept financial responsibility and to invest in environmentally sound technologies by focusing upon the economic benefits to be gained from doing so. A third barrier is the opposition from farmers. As the clean up of the Dianchi Lake clearly illustrates, in a situation where a diverse number of pollution sources are involved, farmers are unwilling to accept their share of responsibility for environmental clean-up because they believe that they are not largely to blame. This attitude is largely reflective of the unequal economic power structures within the community.

Shared responsibility is a normative goal arrived at through a process of information sharing and participation. The former relies upon improved dialogue between government agencies, enterprises and citizens, whereas the latter requires the actual involvement of various stakeholders in environmental planning and investment. On this basis, we can distinguish a fourth, social dimension to the conceptual framework by expanding upon the information requirement.

The earlier discussion has stressed the importance of four key dimensions in developing environmental capacity in China – *financial efficiency, institutional cohesion, technological advance* and *shared responsibility* (see Figure 6.1). Each dimension can be strengthened in isolation from the others. However, all four dimensions are interdependent and therefore form a virtuous circle. A financial mechanism is required to enhance the level of technological advance, which in turn is needed to increase the cost-effectiveness of financial investment.

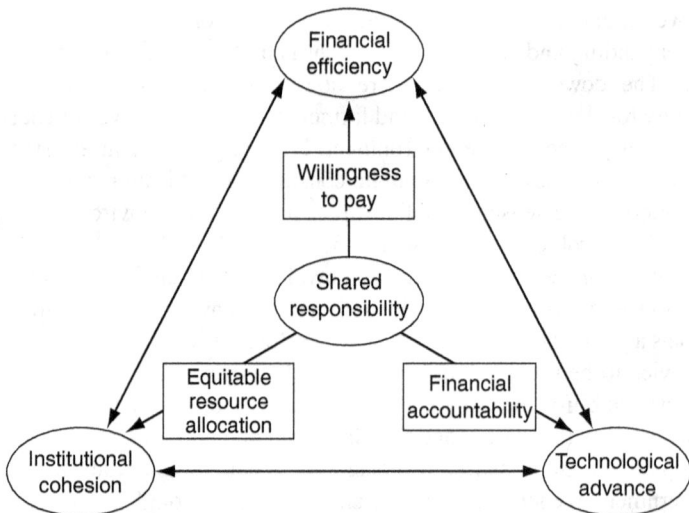

Figure 6.1 A conceptual framework for developing environmental capacity.

Technological advance also facilitates institutional cohesion through improved monitoring procedures and information management. Institutional cohesion, in turn, leads to more effective technological dissemination and financial management. Finally, in closing the virtuous circle, improved interagency coordination can, in turn, increase the effective utilization of financial resources.

Shared responsibility acts as a common thread running through each of the other three dimensions. It also has an important enabling function. For example, enhancing a sense of responsibility amongst stakeholders can lead to an increase in the willingness-to-pay, which in turn builds financial capacity. Stakeholders are far more likely to be willing to pay for environmental quality if they feel that they have a social obligation to do so. Shared responsibility can also strengthen technological capacity by increasing the financial accountability of enterprise managers. Environmental investment in industrial enterprises is more likely to be effective when enterprise managers feel responsible for repaying the loans rather than simply treating them as a gift. In the same vein, shared responsibility can strengthen institutional capacity by improving the equitable allocation of resources. As noted earlier, institutional cohesion amongst government agencies in China is severely constrained by poor environmental planning and budgetary coordination. A participatory mechanism is needed to develop interagency coordination, but to be effective this requires a shared sense of responsibility on the part of the government agencies involved to allocate budgetary resources for the environment on a more equitable basis.

Clearly, one of the major problems involved in operationalizing such a framework is the issue of how to determine when capacities have actually been strengthened. This research relied upon the four key indicators identified in Chapter 1, namely improvements in pricing structures, interagency coordination, the cost-effective use of technologies and monitoring activities. But through the course of this research it became clear that these minimal requirements were not sufficient to gauge the effectiveness of capacity building in the longer term. To this end, more attention needed to be paid to evaluating the degree of supervisory control over environmental loans, the extent of enterprise sharing in cost-effective environmental technologies, and the quality of the interaction between environmental planners and beneficiaries. To operationalize the conceptual framework, Table 6.1 presents a number of evaluative criteria for determining environmental capacity in both the short and longer term.

The strengths and weaknesses of international environmental aid

Despite the centralization of political control and environmental planning in China, environmental donors have had some positive effects upon local environmental capacity. Put simply, by promoting appropriate pricing and cost-recovery polices the World Bank has enhanced financial capacity in Liaoning, Yunnan and Guangxi. The UNDP has been the major driving force behind improved institutional capacity in Shenyang by assisting in the establishment of new panels of representation. It has also been the prime motivator behind technological advance in Benxi, although improvements have largely centred upon technological acquisition rather than the more problematic issue of technological diffusion.

Table 6.1 An operational framework for developing environmental capacity

Short-term indicators	Long-term indicators
Capacity for financial efficiency	
Resource mobilization	Resource utilization
• Efficient pricing policies	• Realistic matching of financial
• Establishment of environmental funds	resources with environmental needs
for enterprise investment	• Long-term financial forecasting
• Adequate sources of capital for the	• Adequate screening procedures and
operation and maintenance of	supervisory control over environmental
environmental services	loans
Capacity for institutional cohesion	
Interagency coordination	Interagency openness
• Establishment of new panels of	• Improved dissemination of ideas and
representation	information disclosure
• Improved strategic planning	• Improved public consultation
• Involvement of senior status officials in	• Involvement of key government
strategic coordination activities	stakeholders in strategic planning and
Capacity for technological advance	implementation
Knowledge creation and acquisition	Knowledge diffusion
• Cost-effective use of technologies	• Information dissemination
• Improved monitoring procedures	• Enterprise sharing of cost-effective
• Increased technical auditing expertise	production processes and products
	• Involvement of enterprise managers
	and workers
Capacity for shared responsibility	
Information sharing	Participation
• Improved dialogue between citizens,	• Involvement of citizens, enterprises and
enterprises and government agencies	government agencies in environmental
• Educational activities	cost-sharing
• Public and enterprise involvement in	• Involvement of local beneficiaries in
information sharing	environmental investment
	• Quality of the interaction between
	environmental planners and
	beneficiaries

Japan's environmental assistance has had a weaker impact upon environmental capacity, although its emphasis upon local ownership has helped to strengthen financial capacity in Liuzhou, Huhhot and Guiyang by encouraging local beneficiaries to seek out the most cost-effective environmental technologies.

These improvements in local capacities have, in turn, led to some initial environmental benefits. Local reports show that pollution loads have decreased considerably in Shenyang, Benxi and Liuzhou. However, it is important to stress here that these reports have not, as yet, been confirmed by the donor agencies. In Shenyang, pollution loads may well have decreased but the main causal factor could have been a reduction in production per se (arising from the large number of 'laid off' workers) rather than a real improvement in pollution control. At a more modest level, perhaps the best example of enhanced environmental quality as a result of capacity building is the UNDP's sustainable city project in Shenyang. For many years, bureaucratic conflict had prevented the city from developing its residential sanitation. As a consequence of the project, improved interagency coordination led almost immediately to the construction of 100 public toilets.

A deeper investigation into the local perceptions of government officials and enterprise managers reveals that the tangible benefits to be gained from capacity building are only part of a much bigger picture. Equally important are the more intangible benefits relating to increased local awareness and understanding. The proposed conceptual framework for developing environmental capacity takes into account the importance of these benefits by focusing upon the notion of shared responsibility.

Facilitating the capacity to share responsibility, however, has been confined to a small number of projects involving a few local communities, government agencies and industrial enterprises. In a few small localities in Guangxi, the World Bank has played an important pioneering role in enhancing the local sense of shared responsibility. This was largely because equal emphasis was placed upon economic efficiency and participation. Local beneficiaries were encouraged to invest in environmental services as well as participate in their design, operation and maintenance. At the elite level, the UNDP project in Shenyang is the best example of how stakeholder participation can enhance interagency responsibility for the environment. And the UNDP project in Benxi also demonstrates how appropriate technological interventions can increase enterprise responsibility towards controlling pollution. But in all the earlier cases, it is important to recognize that facilitating shared responsibility was not the donor's main intent; it was a positive externality that was largely recognized by the beneficiaries rather than the donors.

Understanding the wider parameters of environmental capacity building requires a more comprehensive evaluation of attitudinal change. The World Bank, for example, has been relatively successful in raising awareness of the need for sound financial management amongst local utility companies and government agencies. But given the political and economic stakes involved, both donors and local beneficiaries have been more focused upon practical outcomes than changing attitudes. This is largely because on the donor side physical improvements are far easier to measure than changing attitudes, and on the recipient side, visible improvements in environmental management are critical for winning local political and public support. Although attitudinal change is difficult to measure and not always appreciated by the local recipient, its incorporation into project assessment methods and evaluation is critical and requires ongoing research efforts. Unless new practices, skills and knowledge are embedded within the local consciousness they are unlikely to endure over time.

An analysis of the pitfalls of international environmental aid to China can be equally illuminating. Donor aspirations can often become entangled with local interests in unexpected and sometimes negative ways. A major problem with directing environmental aid towards infrastructure building is that the absorptive capacity for environmental infrastructure in China is huge but the return on investment is typically low. Hence donors risk default on the repayment of loans and/or a misuse of local funds, as in the case of Japanese environmental loans serving *unintentionally* as a subsidy for non-performing SOEs. The potential negative consequence is clear: by stalling the reform process, environmental aid may well be counterproductive to reducing China's emissions in the longer term.

Getting the prices right and developing institutions for environmental services clearly goes beyond the parameters of the traditional technocentric approach to environmental management. In theory, an approach that takes into account financial and institutional factors should have a better chance of success. But in practice, this is no easy task. The difficulties involved in imposing financial discipline upon existing institutions in developing countries are well known. China is no exception: an underlying assumption exists amongst government officials that 'money from one pot or another is all the same thing'. The task with respect to institutional reform in China is even more difficult because it requires creating a market in an area still dominated by socialist allocation.

Too strong an emphasis upon the market also risks undermining local efforts towards implementing cleaner production and eco-agricultural practices. This is particularly evident in the poorer regions of China where local officials are grappling with very complex environmental problems that involve multiple sources and require the participation of farmers who do not necessarily feel any responsibility for environmental protection.

One might easily assume that a focus upon participatory practices could overcome the deficiencies of the traditional and economistic approaches to environmental management. Yet, there does not exist a singular methodology for applying such an approach to differing socio-economic contexts. It is significant that participatory decision-making helped to increase interagency coordination in Shenyang but this was not replicated in Wuhan. In the latter case, progress was constrained by local political structures and sectorally based government agencies. A more fundamental problem with the participatory approach, however, is the lack of attention given to financial efficiency. This can have a negative effect upon capacity building by reinforcing local dependency upon foreign investment. Without the necessary means to implement concrete project outcomes, political commitment is also likely to wane.

From the boardroom to the township: what makes international environmental aid work?

An analysis of the strengths and weaknesses of international environmental aid to China underscores the critical importance of adjusting donor approaches to accord with local socio-economic circumstances. In comparing the effectiveness of the three donor approaches in poor and relatively developed regions of China it becomes clear that no single approach is uniformly more successful than the other alternatives. Instead, it would seem that different approaches work better in particular regional settings (see Table 6.2).

In the case of the Japanese approach, project outcomes relating to capacity building are relatively weak, and more visible in China's poorer regions. This is

Table 6.2 A regional comparison of environmental project outcomes in China[a]

Approaches	Financial efficiency	Institutional cohesion	Technological advance
	Relatively developed[b]	*Relatively developed*	*Relatively developed*
Engineering approach	Low[c]	Low	Low
Market-institutional approach	High	Medium	Low
Human development approach	Low	Medium	High
	Poor regions[d]	*Poor regions*	*Poor regions*
Engineering approach	Medium	Low	Low
Market-institutional approach	Medium	Low	Low
Human development approach	Low	Low	Medium

Notes

a The social dimension of environmental capacity has been excluded from the table because project outcomes have been limited.

b Relatively developed regions include northeast and central China (including the municipalities of Shenyang, Jinzhou, Benxi, Dalian and Wuhan).

c Here the use of low, medium and high refers to the degree of project effectiveness based upon the evaluation criteria presented in the above conceptual framework.

d Poor regions include north and southwest China (including the municipalities of Huhhot, Kunming, Guiyang, Nanning, Liuzhou, Guilin, the county of Yangshuo, and the townships of Wulong, Dounan and Bianyang).

hardly surprising, given that current project goals are narrowly designed to produce good engineering outcomes with little regard for economic and institutional imperatives. The World Bank's approach has been more effective in developing financial than institutional capacities, and project outcomes overall are higher in the relatively developed regions of China. The UNDP approach has also been more successful in the richer regions of China, especially with respect to developing technological capacity. Its effect upon institutional cohesion has been less because of the discrepancy in levels of political commitment between the cities of Shenyang and Wuhan.

Local conditions for facilitating environmental capacity

Why do different donor approaches work better in particular regional settings? The necessary pre-conditions for capacity building involve a number of factors, including the severity of the environmental problem, the political status of the local EPB, the quality of local environmental research institutes, and the degree of public involvement. Above all, the case studies demonstrate the importance of three key economic, political and spatial conditions, each related to one donor approach: *the degree of industrial restructuring* (engineering approach); *the degree of pre-existing political commitment and institutional capacity* (human development approach) and *the degree of urban concentration* (market-institutional approach).

The engineering approach works better in regulated local economies where SOE reform is less advanced. This is because the approach targets pollution control without taking into account the need for cost-recovery policies. While this may be effective in poorer regions where SOEs are likely to remain under government control, it is not economically feasible in the more developed regions of China where these enterprises are under considerable political and economic pressures to reform. In a transitional setting, the need for a financial return on environmental investment is especially important. But this is only possible if the enterprises involved are economically competitive. To provide an incentive, one possible solution is to establish an environmental fund with fairly stringent loan conditions. Recall from Chapter 3 of this book that this is exactly what Japan did in the 1960s when trying to control its own industrial pollution. It set up an environmental pollution fund – the JEC. Yet, ironically it is the World Bank that is putting this experience into practice in China. While the Bank's environmental fund in Liaoning is by no means a perfect solution because of the problem of weak supervisory control, it makes far more sense than sinking funds into failed enterprises.

In regions with higher levels of economic development, it is also not clear that concessional loans for point source control are a better option than private sector investment. Japanese private firms are increasingly interested in investing in pollution abatement in China. The major problem with foreign aid, as noted by local officials in Shenyang, Benxi and Dalian, is that the time lag involved between the application and actual disbursement of environmental loans is not suited to the current pace of industrial reform. In poorer settings, the engineering approach is likely to be more effective simply because the opportunities for private investment are more remote and environmental funds in general are less readily available. The size of investment per se can stimulate political commitment, an important prerequisite for capacity building, as well as provide an important financial incentive for local industries to take environmental protection more seriously.

This is not to suggest that Japan should simply focus its attention upon infrastructure building in China's poorer regions. As this study clearly illustrates, Japanese loans may be effective in reducing pollution loads but in the longer term they are likely to have an adverse effect by reducing the incentive to invest in managerial and institutional capacities. If engineering is to be the main objective, a more sustainable solution would be to link point source control with innovative process improvements and broad based stakeholder participation.

The human development approach is highly dependent upon pre-existing political commitment. Although to some extent this is a necessary condition for all approaches to environmental management, it is particularly important in the case of human related development goals. This is because local Chinese officials are less familiar with the concepts of stakeholder participation and cleaner production than they are with engineering and market efficiency. The approach, therefore, requires more political consensus building, both amongst local political elites and government bureaucrats, before it can be effectively implemented.

A minimal level of pre-existing institutional capacity is also important. In China, political status usually determines institutional capacity but it is also rooted in access to resources, the quality of staff[3] and the quality of the supporting intellectual infrastructure. Above all, the major difference between the EPBs located in the poor regions of China and those located in the more developed regions lies in the fact that the latter have a stronger strategic capacity to integrate environmental problems into development concerns. This, in turn, relies upon strong intellectual support, especially in relation to bridging the divide between environmental science and environmental economics. Ironically, the need for pre-existing capacity means that the human development approach is likely to fail in poorer regions where it is arguably most needed. These regions are doubly disadvantaged because the approach does not place enough attention upon improving financial efficiency, and therefore risks undermining political commitment and capacity building in the longer term. To be more sustainable, the intrinsic value of participation needs to be reconciled with the instrumental need for economic viability.

As noted earlier, the market-institutional approach is also less effective in poorer settings. But contrary to many expectations in the literature, this is not because local beneficiaries are either unwilling or unable to pay. It is clear that, under the guidance of the World Bank, appropriate pricing has been implemented in China regardless of the level of economic development. Between 1994 and 1998 water tariffs doubled in the poorer cities of Nanning and Guilin and tripled in Kunming. In comparison, although tariffs also doubled in Dalian, they increased by less than 30 per cent in Shenyang (see Table 6.3). This seems to vindicate the Bank's position, outlined in Chapter 5, that the 'willingness-to-pay' for water is relatively high amongst poor communities.

The real problem for the World Bank is that a sole emphasis upon market institutions is less appropriate in poorer settings in China where the urban/rural divide is more ambiguous. Many former rural villages on the periphery of major cities have evolved into semi-urban settlements that currently fall outside traditional administrative boundaries. Consequently, an approach that only targets market incentives is not likely to work. A low degree of urban concentration calls for higher levels of participation, both between farmers, industry and city residents and between agricultural, planning and government agencies. In poor semi-urban settings the key to effective environmental management lies in striking a balance between economic incentives and stakeholder participation. It is also important to recognise that no matter how much genuine participation takes place, few

Table 6.3 Water tariff adjustments in China (RMB m³)

Years	Poorer cities			Relatively developed cities	
	Nanning	*Guilin*	*Kunming*	*Shenyang*	*Dalian*
1994	0.20	0.20	0.32	1.10	1.20 (0.18)
1995	0.32	0.32	0.60 (0.18)	—	—
1998	0.40 (0.12)	0.42 (0.18)	0.95 (0.35)	1.50 (0.50)	2.50 (0.80)

Source: Compiled from project data and interviews.

Note
Figures in brackets represent the amount included for wastewater charges.

farmers in China (as in many other developing countries) would be convinced that they bore responsibility for environmental clean-up equal to that of industry and city residents. Their position is justifiable. And under these circumstances, more attention needs to be paid towards introducing eco-agricultural practices and cleaner production technologies.

An alternative approach in poorer settings

We have seen that to date donors have made little effort to adjust their approaches to accord with diverse social and economic circumstances. The only exception is the World Bank's small area improvement projects in Guangxi Zhuang Autonomous Region, that take into account local realities and also provide an exemplar of how to reconcile incentives with participation in poorer settings. This alternative approach can best be described as one of *participatory ownership* in that ownership by local users and stakeholders is the desired outcome and participation the practical means of achieving it.

The iterative process involved means that project outcomes take longer to materialize. As an isolated case, it is difficult to determine the scope for further replication in China. However, based on the project investigations, four observations suggest that the approach could be effectively applied to county/township and semi-urban settings. First, relinquishing power is more difficult for a municipal government than a county government because traditionally in China the lower echelons of government have enjoyed more autonomy and political flexibility. Second, social organization in county and township settings tends to be looser and more receptive to change. Third, township participation can be managed as a single unit whereas neighbourhood participation within larger municipalities requires more compromise and is often at odds with the larger system as a whole. And fourth, the notion of 'representative participation' is becoming more widely accepted amongst township and county government agencies in the wake of recent village elections. It is, perhaps, not mere coincidence that some of the first democratic village elections in China took place in Guangxi. At the time of my visit, an election was about to be held in the township of Baisha where one of the Bank's small area improvement projects was located.

In representing a synthesis of economic and societal perspectives on environmental management, the *participatory ownership* approach is arguably more likely to build sustainable local capacity. The problem is that such an approach can only be applied on a small scale in the appropriate political context. Few precedents exist and, therefore, the approach requires a considerable amount of time and patience on the part of donors and local implementing agencies. An additional difficulty is that genuine participation evokes a different image for international donors than it does for local beneficiaries. As the projects in Nanning and Yangshuo attest, the idea of participation can also take on different meanings amongst the beneficiaries involved.

To summarize, to a limited extent, environmental donors have been successful in facilitating local environmental capacity in China. Taking into account the relationship between particular donor approaches and local conditions could further enhance aid effectiveness. Yet, in order to build sustainable environmental capacity, donors need to recognize the important linkages between engineering, market and participatory approaches to environmental management. Three questions are relevant here. First, does the commitment exist within donor institutions for improving environmental assistance? Second, to what extent do donors have the institutional capacities to do so? And third, what are the prospects for improving inter-donor coordination?

Donor constraints

International environmental aid to China is clearly motivated in part by mutual self-interest. Both donor and recipient stand to gain politically, economically and diplomatically from investing in environmental protection in China. But it is not entirely clear the extent to which donors are committed towards assisting China in the longer term, or indeed, whether they are motivated to ensure effective outcomes on the ground. It is also doubtful whether donors have the necessary institutional capacities to expand and improve upon their environmental aid programmes to China.

For Japan, the threat of acid rain provides a long-term incentive to promote effective environmental management in China. But is this motive sufficient? Addressing the acid rain problem in China and thereby reducing the potential threat to Japan would be in the interests of the average Japanese taxpayer. Furthermore, an environmental aid programme that targeted the more impoverished areas in China would be more likely to meet with international approval. For this to succeed, Japan would need to work more closely at the local level, which in turn would require strengthening its cooperation with local community groups. In 2003 the Japanese Foreign Ministry announced a new plan to introduce local government ODA.[4] However, funding to NGOs still accounts for only a very small percentage of the ODA budget (0.5 per cent in 2002 compared to more than 30 per cent of the US aid budget).[5]

The continuous challenge that the Japanese government faces is in trying to reconcile public demands for more effective development assistance with the demands of the private sector long accustomed to the protective blanket of ODA.

Japanese firms are becoming more assertive in their relationship with aid policy-makers in Tokyo. Obviously, commercial interests would not be satisfied if JBIC were to concentrate its environmental loans in the poorer regions of China, where markets are relatively underdeveloped and basic infrastructure weak or non-existent. On the other hand, if loans became more tied to Japanese firms no capacity building would take place: Japanese manufacturing companies have a limited incentive to promote localization because their motive is simply to participate in demonstration projects to gain experience for future export plans.[6] One possible way forward for the Japanese would be to limit commercial interests to METI's Green Aid Plan, which would be far more likely to succeed on a commercial basis if it were concentrated in the northeast of China where commercial linkages between China and Japan are already well established.

It is also important to consider Japan's political and diplomatic interests, which for geographic and historic reasons are equally inextricably linked to China's northeast. An unknown factor is the extent to which Japan would be willing to provide environmental loans to factories in the northeast on the basis of historic responsibility, and regardless of economic viability and the risk of generating negative environmental outcomes. It is significant that many of the factories in Liaoning province that are currently receiving environmental aid from Japan were formerly owned by Japanese companies.

For the World Bank, the future of environmental aid to China lies in the extent to which local beneficiaries are able to secure a return on investment. This, of course, is more likely to happen in China's richer regions, as illustrated by the Bank's environmental project in Shanghai. Before the project was implemented, the rivers in Shanghai acted as the city's main sewers but now, with sound cost-recovery policies and strong market institutions, Shanghai has a fully functioning wastewater management system.[7] In time, there does not appear to be any reason why the cities of Shenyang, Benxi and Dalian should not follow suit. The problem is that by narrowly focusing upon municipal environmental infrastructure in more developed urban settings the Bank would be forfeiting an opportunity to play a more central, and indeed, influential role in environmental management in China. To this end, the Bank needs to address the difficult issue of industrial pollution control in poorer regions without compromising its interest in promoting SOE reform.

Clearly, the UNDP is less concerned with securing an economic return on investment. It is more interested in enhancing its influence beyond Beijing. This is most likely to occur in China's industrial pollution heartland in the northeast where urban pollution problems are most severe. Greater visibility and stronger project outcomes, in turn, are more likely to attract additional donor funding. But instead of concentrating its efforts on a small number of locations in order to enhance the benefits of small-scale investment, the UNDP is moving in the opposite direction towards a wider dissemination of demonstration projects across China. In the longer term, it is difficult to see how the UNDP can seriously maintain a commitment towards capacity building, let alone expand its presence, when its resources are spread so thinly.

For all three donors, perhaps one of the biggest constraints upon a long-term commitment towards addressing environmental problems in China is the issue of 'aid overstretch'. At a time when aggregate aid investment in environmental management is beginning to make a difference on the ground, aid institutions are already shifting their attention towards other priorities such as gender, good governance and poverty alleviation. While these priorities are all clearly laudable goals, and indeed have important linkages with environmental protection, it is difficult to see how aid can be made more effective when the goal posts are constantly changing and when policy ambitions far outstrip the capacities of donor institutions.

Institutional weaknesses

A second major barrier to improving environmental aid to China is the lack of institutional capacity within the donor organizations themselves. Implementation on the ground, albeit successful in part, has fallen far short of the ambitions of donor institutions in Tokyo, Washington and New York; not only because expectations were not realistically matched with local conditions but also because of serious weaknesses on the part of donors. Aid agencies are aware of their own shortcomings, but they seem unable to address them because of deeply entrenched institutional constraints.

Japan has yet to establish an integrated framework for providing environmental aid to China. In retrospect, the Japanese have largely reacted to rising environmental pressures in China in an *ad hoc* manner without any clear direction or strategic vision. This is aptly summed up in the words of one of the managing directors at JBIC:

> If we review the process by which action was taken, we see that responses were in part rushed due to the mounting pressure of global environmental problems...we all feel we are putting efforts into solving environmental problems, but we need to be concerned that the substance of the programmes keeps up with the spirit.[8]

The UNDP and the World Bank are equally not immune from self-criticism. Both agencies have been under political pressure to act immediately, especially in light of positive policy signals in Beijing. As a result, projects have often been rushed through with little attention given to actual outcomes. In the words of one official at the World Bank: 'We are churning these projects out like a cookie machine.'[9]

Coordination amongst Japanese agencies over environmental assistance is also weak. METI, JBIC and JICA all organize separate missions to China with different agendas. According to one JICA official, the Chinese side is now starting to lose patience and refuse requests for Japanese aid missions.[10] More recently, Japan has moved a step in the right direction by establishing expert committees composed of JICA and JBIC officials. Nevertheless, a rigid divide remains between loans and technical assistance that acts as a serious impediment to transferring the Japanese experience in cleaner production to China. It is precisely this divide that reduces the overall effectiveness of Japanese environmental aid.

Contrary to views expressed in Tokyo, the problem of developing technological capacity in China is not simply a question of local non-compliance; many of the local EPBs and enterprises involved in Japanese environmental projects positively supported the idea of cleaner production. Instead, the fundamental problem lies with bureaucratic resistance in Tokyo and the divide between loans and technical assistance. In Japan's first aid study group report on the environment in 1988, the chairperson, Dr Michio Hashimoto, concluded:

> It is essential that the cooperative relationship between the two organisations, the Japan International Cooperation Agency and the Overseas Economic Cooperation Fund, be strengthened even more so that environmental considerations can be carried out consistently.[11]

Fifteen years on this structural divide between Japan's two agencies responsible for aid implementation is the single biggest obstacle to Japan's long-term contribution towards strengthening China's environmental capacity.

As a consequence of the merger in 1999 between the OECF and the Japanese Export and Import Bank to form JBIC, the future prospects for strengthening the link between infrastructure and technical assistance do not look bright. The merger reflects the pro-loan stance of the Japanese aid bureaucracy. With attention now focused upon reconciling Japan's international financial operations (including Japanese export/import promotion, Japanese investment activities overseas and the stability of international financial order) with its overseas economic cooperation activities, it is difficult to see how technical assistance will become a more important part of the aid agenda. Improved donor coordination is also likely to be constrained by the limited number of staff working at JBIC. In fiscal 1999, JBIC's financial commitments reached almost ¥3 trillion (approximately US$27 billion), equivalent to the US$29 billion (including IDA credits) committed by the World Bank.[12] But whereas the World Bank had a staff of over 6,000 professionals to manage these funds, JBIC had a staff of only 889.[13]

In striking contrast to the lack of vision within the Japanese aid bureaucracy, the fundamental problem for the UNDP is that it is somewhat overwhelmed by its own compellingly visionary agenda. Although the UNDP stands out amongst other bilateral and multilateral donors in its innovative stance, the UNDP's forward thinking often lacks institutional cohesion. This is certainly the case in relation to the UNDP's participatory approach, which is largely driven by the organization's younger and more motivated recruits. In the words of one UNDP official: 'Old school bureaucrats are not used to the participatory approach. No career incentives or performance-based measures exist to encourage UNDP officials to implement participatory approaches.'[14] Decentralization of the UNDP's activities also makes it difficult to enforce new ideas and approaches upon country programmes that, at least in the case of China, are largely driven by recipient concerns. The lack of institutional cohesion is even apparent in relation to the UNDP's efforts to promote global environmental programmes. Although the GEF has greatly enhanced the UNDP's financial capacity for environmental aid, according to the GEF regional coordinator for climate change: 'within the UNDP itself, the GEF has not been properly assimilated into the mainstream. Only now is a greater consciousness taking place.'[15]

A second institutional constraint upon improving environmental aid to China is the weak link between the UNDP's policy dialogue and its demonstration projects. As mentioned in Chapter 4, the UNDP's main objective is to enhance policy influence at the national level in developing countries in order to complement its development activities at the grassroots. But in China, the UNDP needs to shift in the opposite direction. Its environmental policy towards China over the past decade has been to reinforce political commitment at the national level. Consequently, its capacity building efforts lack the necessary resources and support networks with local officials and enterprises. In some cases, the responsibility for managing projects is in the hands of Chinese institutions in Beijing, and UNDP project managers have little idea about what is actually happening on the ground. This dilemma is stimulating a debate amongst UN organizations in Beijing as to whether representative offices should be located nearer local beneficiaries whose interests the UN agencies are seeking to serve.[16]

Like the UNDP, the World Bank also lacks the necessary institutional flexibility to address its weaknesses in relation to environmental aid. Although the Bank is moving in the direction of 'learning by doing' rather than 'doing without learning', too much emphasis is still put upon project preparation at the expense of implementation, and Bank procedures are complex and time consuming. As an illustration of this point, compare the World Bank project in Liaoning with the UNDP project in Benxi. The former had a project office with 25 permanent staff and a total budget of US$338 million, whereas the latter had a project office with 5 staff and a total budget of US$530,000. Relative to investment in resources, the UNDP's project was more effective; factories in Benxi were already recouping some of their investment in pollution control. One could argue that the Bank's investment was particularly high because of its involvement in developing municipal infrastructure. But Japan seemed to be fairly successful in building environmental infrastructure with far fewer people. The Japanese project office in Liuzhou, for example, had a small staff of six people and a total budget of US$150 million. One can only assume that the staff working in the Bank's project office in Liaoning were largely preoccupied with project administration rather than implementation.

As illustrated in Chapter 5, an additional problem is that few career incentives exist within the Bank for promoting innovative practices. Recent reforms appear to have increased rather than resolved institutional tensions. On the one hand, various social, economic or environmental networks have been created to promote greater collaborative efforts across sectors and disciplines. On the other, the Bank has become more like the private sector by creating an internal market within the institution for economists and technical experts to compete to sell their services to loan managers.[17] This, of course, poses a problem for some environmental services such as air pollution that do not have market-determined prices. In such an institutional climate, participation specialists also risk overselling their expertise. It is interesting that the task manager responsible for designing the Bank's small area improvement projects in Guangxi found himself caught between two different camps: on one side, the Bank's participatory experts anxious to sell their services; and on the other, his more orthodox colleagues who were deeply sceptical of institutional innovation. In his words:

> I was confronted by an arsenal of poverty alleviation and participatory experts who advised me that 'reducing poverty' or 'participation' should be an end in itself. This was not the goal. The goal was to use participation as a means of facilitating sustainable project outcomes at the community level. At the same time, more orthodox task managers and task team members found the process to be too frustrating. In the World Bank, professional pride is strong and many task managers get caught up in fiduciary concerns – checking technical reports, project proposals, approvals etc. without having the time to think about development concerns.[18]

Despite the fact that the Bank is putting pressure upon the unit responsible for environmental loans to China to be more innovative, it is unlikely that the approach will become an integral part of the World Bank's environmental assistance to China for a number of reasons. First, although task managers in the urban environment unit of the Bank have been willing to overlook some of the necessary loan procedures in the interests of promoting project innovation, their more fiduciary counterparts in the loans disbursement division have not been so obliging.

Second, such a demand-driven approach is likely to create conflicts within the Bank between urban and rural sectors, engineering and social disciplines, and participatory and market reform ideologies. Third, in order to build confidence and trust on the ground, the approach requires sustained personal commitment that is at odds with the Bank's high turnover of project consultants. This can be illustrated by comparing the UNDP's sustainable city projects in Shenyang and Wuhan with the Bank's small area improvement projects in Guangxi. In the former case, the same project coordinator visited the project sites three times a year over the duration of the project, whereas in the latter case, a different consultant visited the project site once a year. Fourth, without available IDA credits in China, task managers will have to make greater efforts to solicit grant funding from other donors. Hence, not only the World Bank but also a number of other donor institutions will have to be convinced of the potential benefits of investing in the approach. Taking into account all of these factors, the Bank is likely to conclude that for such a small amount of funds it is not really worth it. This study is full of ironies and paradoxes. One of the most obvious is that the World Bank, with its advantage of economic largesse, is taking a bottom-up approach to participation in China whereas the UNDP with its comparative advantage in smallness is taking a top-down approach.

The donor coordination problem

The question still remains as to whether one donor institution could incorporate an integrated environmental approach into its mainstream operations or whether inter-donor coordination would be more effective. A growing convergence of principles and approaches now exists amongst donors at the policy level. Convergence in practice, however, would require considerable institutional adjustment. Interagency coordination is less radical but current efforts in China have not been particularly promising: donors have a tendency to protect their own territory, and different corporate cultures and methodologies act as a major impediment to increasing cooperative efforts. Recall that very little donor coordination was taking place between the UNDP, European Union and the World Bank over urban environmental planning in Shenyang or between the Japanese and the UNDP over acid rain control in Guiyang.

The emphasis upon partnerships for development following the Millennium Summit in 2000 could help to alleviate the donor coordination problem. Of particular relevance is the new cost sharing arrangement between the Department for International Development in the United Kingdom and the World Bank to provide concessional funding for social sector and poverty alleviation projects in China. In 2003, three projects were already underway relating to health (tuberculosis control) and basic education. Yet, realistically such kinds of partnerships can mask historic rivalries. Indeed, the deep-seated competition between the World Bank and the UNDP is particularly evident in relation to capacity building. According to the project coordinator of the UNDP sustainable city projects in China:

> The World Bank is putting more emphasis upon capacity building but many different consultants are used and the main objective is still a return on investment. The World Bank has adapted the sustainable city concept and procedures to its own model of global cities programmes with clearly defined time constraints and outcomes – this misses the point because the process itself is crucial for capacity building and success must be flexibly defined.[19]

In light of the continuing turf wars between donors, it is little wonder that the UNDP has only marginally succeeded in its self-proclaimed role as a coordinator and facilitator of international environmental aid to China, especially in view of the fact that it also has its own turf to protect. A better solution would be for donors to encourage recipient countries to build their own capacity for coordinating development assistance. Research shows that a centralized agency in the recipient country for coordinating all aspects of development assistance can considerably improve donor coordination.[20] This would allow the UNDP to concentrate its efforts on coordinating capacity building. Donor coordination is a perennial problem but it is more likely to succeed if it is targeted to one main objective.

Without improved inter-donor coordination, environmental aid is likely to be far less effective. In the future, it is vital that donors cooperate over stakeholder participation and cleaner production in order to raise their status in the eyes of local Chinese officials and enterprise managers. Moreover, donors need to cooperate over the delivery of environmental aid in order to reduce the problem of cost-recovery. From the recipient perspective, international environmental aid is one area where competition is not always useful.

To conclude, it is now abundantly clear that the time has come for donors to redefine their role in providing environmental aid to China; the fragmented approach has reached its limit. This not only requires a more in-depth understanding of local contexts but also considerable innovation and adjustment on the part of donor institutions. In particular, environmental loans and grants need to be coordinated, the sectoral divide between urban and rural environmental assistance needs to be reassessed in light of the growing ambiguity in China, and inter-donor coordination should be realigned to target capacity building. Above all, environmental aid needs to be driven by implementation concerns rather than ideologically driven policy agendas or the unwieldy processes of project administration. In the absence of radical moves in the direction of greater institutional flexibility and inter-donor coordination, the future prospects for effective environmental aid to China are far less promising.

Conclusion

An initial expectation at the beginning of this study was that entrenched local political and economic interests would seriously impede any attempts by international donors to facilitate local environmental capacity in China. In the existing scholarship on environmental management, government officials at the local level are not seen as having strong incentives to implement environmental measures because economic imperatives heavily outweigh environmental concerns. A second expectation was that both market-oriented and participatory approaches to environmental management would be difficult, if not impossible, to implement in light of China's state-centric focus upon engineering solutions and environmental regulations. This book has demonstrated that both of these expectations are open to challenge. Although the problem of non-compliance at the local level remains, it is by no means uniform across all regions of China. Moreover, at both the central and local government levels, Chinese officials are increasingly willing to experiment with non-state approaches to environmental management.

It would seem that recipient commitment is less of a problem for donors than actually understanding the complexities involved in building environmental capacity. The challenges are multiple: to control industrial pollution while supporting the reform of SOEs; to invest in environmental infrastructure while ensuring that the institutional arrangements are in place to achieve a return on investment; to promote institutional cohesion while reconciling strong vested interests; to encourage citizen involvement while maintaining the support of political elites and to overcome psychological barriers which reflect deep-seated traditions and entrenched ways of thinking. Efforts to strengthen environmental capacity must be grounded in an in-depth understanding of China's development needs, political interests, institutional practices and social traditions and attitudes.

Despite the complexity of the task involved, donors have had some success. The UNDP has helped to facilitate institutional and technological capacities, and the World Bank has been effective in enhancing financial efficiency. By strengthening ownership, Japan has also had a more indirect effect upon building financial capacity. However, achievements in developing capacity that are likely to endure over time, and which will therefore lead to better environmental outcomes, are far less visible. Positive sustainable outcomes have been constrained in two important ways. First, each donor's approach to environmental management has only had a partial effect upon building environmental capacity. Simply put, the Japanese and UNDP approaches lack economic viability while the World Bank's pro-market approach overlooks critical social, and technological concerns. I argue in this book that environmental capacity encompasses four interrelated dimensions: *financial efficiency, institutional cohesion, technological advance and shared responsibility.* If the goal is to strengthen capacity beyond the instrumental purpose of a particular environmental project, then all four dimensions need to be developed simultaneously.

The second limitation is that donors have pursued the same approaches to environmental management regardless of differing local conditions. No single approach has been equally effective across different regions. Regional variation has been largely determined by three key economic, political and spatial factors relating to each approach; namely, the degree of industrial restructuring (engineering), the degree of pre-existing political commitment and institutional capacity (human development) and the degree of urban concentration (market-institutional). On balance, environmental aid to China has been more effective in China's more developed regions.

The imperfect fit between donor approaches and local conditions is a familiar story in the history of international development assistance and is, therefore, hardly surprising. Nevertheless, this failing has particular resonance in light of recent policy changes in China and at the international level. A central priority of China's tenth Five-Year Plan (2001–2005) is to reduce regional inequities by developing China's western and inland provinces where the largest proportion of the country's estimated 200 million poor reside. These regions are also dispro-portionately affected by poor environmental conditions. Hence the current policy of the Chinese government is to encourage more environmental aid to shift inland. This policy fits squarely with the new development agenda at the interna-tional level that seeks to address the linkages between environmental protection and poverty alleviation.[1] Indeed, the World Bank's new Country Assistance Strategy for China (2003–2005) aims to address both economic inequality and environmental decline.[2] It is expected that approximately 75 per cent of the IBRD loans (total US$1.2–1.3 billion annually) will be disbursed within China's inland provinces.[3] Japan has followed a similar trend. In fiscal 2001, 85 per cent of the JBIC loans were earmarked for inland provinces (4 out of the 13 projects were environmental).[4] If environmental aid is to succeed in the poorer and ecologically fragile regions of western China it will require innovative and sustainable interventions on the part of donor agencies that are carefully attuned to local practices.

This book provides three major conclusions. First, it demonstrates that a local commitment towards the environment does actually exist in China with many local governments now recognizing the limitations of a purely state-centric approach to environmental management. Second, from a theoretical perspective, it shows that a pluralistic approach to environmental management is more likely to achieve positive outcomes in relation to capacity building. The ideological divide that exists between market proponents and participatory advocates has little resonance in practice. Even in a market transitional and authoritarian state such as China, both approaches are necessary to enhance environmental capacity and, indeed, are interdependent. Third, from a policy perspective, the book suggests that urgent improvements are needed in the design and delivery of international environmental aid. To this end, donors should pay equal attention to adjusting their approaches to local circumstances and to addressing the weaknesses within their own institutions. Without these dual efforts, international aid is likely to fail its task of helping to improve environmental conditions in China and it could even lead to negative consequences in the longer term.

The shifting grounds of local environmental management

Local EPBs in China are often characterized as politically weak, largely impotent in regard to enforcing widespread environmental regulations, and sometimes in collusion with the very enterprises that they are supposed to be regulating. The findings in this study convey a counter impression of hard work, commitment and accomplishment. Although local EPBs are still weak relative to production-oriented agencies, many local officials (not only from the EPBs) were found to be indefatigable in pursuing the task of protecting the environment for the benefit of their communities. Cynically, positive environmental action at the local level could be interpreted as keeping up appearances for the sake of soliciting external funding. But a more balanced explanation rests on the premise that a growing political commitment towards environmental protection has now permeated beyond Beijing.

At all 13 project sites visited for the purposes of this study, local non-compliance was not the overriding constraint upon environmental protection in China. The issue of local capacity was equally important, if not more so. Moreover, non-compliance was not always related to local vested interests. The Japanese environmental project in Shenyang, for example, had encountered strong resistance because central government officials in Beijing owned the state enterprises involved. The local constraints upon implementing environmental measures are, therefore, far more complex than a simple disjuncture between a growing consensus in Beijing and dissent within the provinces.

Local political commitment often reflects the severity of the environmental problem; it can also be motivated by the potential for environmental investment, the need to promote tourism, or by national pride in achieving environmental standards. International environmental assistance can serve as a stimulus in two ways: by simply providing funds, or by appealing to the local leaders' sense of pride. The UNDP sustainable city project in Shenyang provides a clear illustration of how the city mayor, Mu Suixin, gave his personal support to the project because it symbolized an important turning point in the city's international reputation. In effect, by implementing the project, Shenyang was moving away from keeping company with the world's dirtiest cities and towards an association with the developing world's more sustainable cities. This suggests that social incentives can be as important as economic incentives in stimulating political commitment towards the environment.

From a broader perspective, it is clear that China is moving away from the previously dominant state-centric paradigm of environmental protection and towards a more comprehensive paradigm of environmental management that embraces market incentives and stakeholder participation. We saw in Chapter 2 that this shift is already beginning to take place amongst government agencies in Beijing that are at least willing to experiment with alternative approaches. The findings in the case studies also demonstrate that a parallel shift is now taking place at the local level in recognition of the limitations of environmental regulations and point source control. In both richer and poorer regions of China, local environment officials are increasingly aware of the need for cleaner production techniques, appropriate pricing policies and interagency coordination.

Although, in general, local activities comply with national policies and regulatory standards, top-down environmental planning is not absolute. Many local governments are taking their own initiatives *vis-à-vis* environmental protection; some government officials in Shanghai and Guangzhou are more forward thinking in this respect than their Beijing counterparts. Unlike government officials in Beijing, local officials are not so caught up in the need for state-of-the-art high technology transfers because they recognize the importance of acquiring cost-effective and appropriate environmental technologies. They are also more open to new ideas in relation to non-point source control, strategic planning, and even participatory practices if centred primarily within government institutions. As illustrated in Chapter 4, the local response to the notion of participation was more one of adjustment rather than obfuscation. Local officials aimed to fine-tune the participatory approach to accord with local traditions and socio-political practices.

As a consequence of the rise in local commitment to the environment, the issue of central control over donor environmental funding is less of a problem than one might first expect. Funds are going where they are most needed and cost sharing arrangements help to ensure compliance on the part of the beneficiaries. Given that environmental problems at the local level are far more immediate and palpable than accounts generated in Beijing, local officials are more aware of capacity needs. Growing local government concern over the lacuna in environmental planning *vis-à-vis* semi-rural townships or satellite towns (*weixing chengzhen*) is a good illustration of how the contours of environmental management are shaped differently in the eyes of local officials compared with their counterparts in Beijing.

However, despite these positive signs, at both the central and local government levels in China capacity building is still not considered to be as important as soliciting external funding for environmental infrastructure. The SEPA in Beijing is preoccupied with attracting funding to implement the central government's Transcentury Green Plan. And at the local level, officials tend to be more concerned with achieving visible outcomes in relation to environmental infrastructure rather than strengthening institutional capacity over the longer term. This is particularly evident in the poorer regions of China where cities tend to depend wholly upon attracting foreign investment.

Beyond the theoretical impasse

By focusing upon environmental capacity, this book has also provided some interesting theoretical insights into the management of environmental problems in developing countries. In the China context, the major problems involved in the application of market and participatory approaches to environmental management differ from certain expectations in the literature. Given the low levels of per capita income and education available in developing countries, the market approach is often criticized as being too technocentric and north-biased. But currently, local governments in China including those in poorer regions are prepared to increase utility prices over a period of time, which essentially vindicates the World Bank's universalistic emphasis upon price incentives. From a local perspective, the average citizen in China stands to benefit considerably from the reform of inefficient state-owned utilities. At present, residents in Nanning, for example, do not have access to water for up to one week at a time, and very often hot water is reserved for local officials. Hence, while market solutions are no panacea, they clearly make an important contribution to solving environmental problems and meeting the needs of ordinary citizens. The difficulty is not price reform per se, for it can be adjusted according to the level of economic development, but the issue of how to build the necessary institutional support for operating and maintaining environmental services at the same time.

The main weakness with the market approach is its assumption that assigning private property rights is a cure for all ills. In most instances, the environment is a public good. The approach is, therefore, limited in regard to solving environmental problems such as air pollution that do not have market-determined prices. Tradable permits could be applied but in order to reduce pollution loads, technological advance is also necessary. Moreover, as illustrated by the World Bank's environmental project in Yunnan, the market approach is not sufficient to deal with environmental problems that involve a wide diversity of pollution sources. Under these circumstances, stakeholder participation is essential.

The participatory approach is often criticized by scholars for being time consuming and less applicable, if not impossible, in authoritarian contexts. While such criticism is justifiable, a more serious problem with the approach is that it often overlooks the critical importance of economic viability. Over time, if participation fails to promote economic interests in addition to political or social interests, it may be rejected. However, it is also important to recognize that, in some cases, the benefits of the approach can outweigh the weaknesses. In China, despite institutional and political constraints, the approach has been equally important in promoting interagency coordination at the elite level and in encouraging a willingness-to-pay at the community level. The strength of the participatory approach lies in its malleability on the ground, and the fact that participatory practices can be applied in diverse contexts regardless of political and socio-economic circumstances.

In situations where some groups are particularly resistant to participating and investing in environmental management, as a consequence of economic power asymmetries within the stakeholder community, developing technological capacity could help to increase commitment towards environmental improvement. For example, in the case of the Dianchi Lake, no matter how much participation takes place those farmers living around the lake would be unlikely to accept responsibility for environmental clean-up equal to that of industry and residents in Kunming. Equally, an increase in fertilizer prices would still not provide a strong enough incentive for farmers to reduce their use of fertilizers and return to more traditional and ecological ways. Furthermore, it would place an unfair burden on poorer farming communities located in more remote parts of the province.

Based upon his investigations in poor rural communities in China's southwest, Jonathan Unger has made the important observation that chemical fertilizer use is a critical means of obtaining grain yields from fields that lack sufficient nutrients. In his words, 'it is the difference between development and hunger.'[5] When livelihoods are at stake, a better solution would be to introduce farmers to the benefits of using eco-agricultural practices, especially in relation to achieving an optimum balance between chemical and organic inputs. If successful, technological capacity building could also help to stimulate an attitudinal change in favour of the environment.

From the perspective of implementation, the theoretical debate over the importance of economic incentives relative to participation and vice versa really misses the point. These perspectives are interdependent because they are grounded in the two fundamental principles of economic viability and shared responsibility upon which effective environmental management depends. The main theoretical conclusion of this study is that if our understanding of solving environmental problems is to progress beyond the parochialism of disciplinary and ideological boundaries, more attention needs to be given to theories that link incentives with participation and technological advance without ignoring the importance of shared responsibility on the part of the stakeholders involved.

It is also important to stress that the role of the state cannot be dismissed, even when it has been historically responsible for causing gross environmental damage. In the case of China, central coordinating efforts are a critical means of mediating conflicts between rich and poorer regions, urban and rural sectors and upstream and downstream communities. In short, effective environmental management demands unprecedented levels of cooperation between various stakeholders that can only be realized by striking a balance between state action, market reform and citizen empowerment. Hence, if progress is to be made in dealing with environmental problems a pluralistic approach is essential.

Future prospects for developing environmental capacity in China

Environmental capacity building is an inherently complex issue. It cannot be seen in isolation from the broader socio-economic context of the recipient country, or the political and institutional interests within the donor institutions involved. The biggest constraint is still the most obvious one: that environmental capacity is poorly understood and recognized on the part of the donor and the beneficiaries alike. It would be tempting to suggest that donors should simply make it a central feature of their policy dialogue with recipient countries. Indeed, developing environmental capacity could become a new form of conditionality. But little evidence exists that policy-based lending actually encourages recipient countries to undertake reforms.[6] As illustrated in all three case studies, the Chinese government was willing to accept environmental assistance and make the necessary policy adjustments because it had a prior commitment to do so.

No amount of lecturing the Chinese on the need for capacity building (i.e. assuming that the donors themselves are committed) would truly make a difference. The message in this book is clear: donors need to *demonstrate* the importance of capacity building by ensuring effective outcomes on the ground. To this end, the international donor community needs to adopt a coherent vision of environmental capacity that is based upon a clear understanding of the key determinants of success. This is no small task given the complexities involved. And, to date, little evaluative work has been carried out. As a starting point, we can distinguish four major lessons from this study.

First, *access to funds on a continuing basis is a key variable for the long-term success of environmental capacity building in China*. Many studies have stressed the need for recurrent financing on a long-term basis in order to sustain capacity building efforts.[7] In the case of the environment this is especially important because of the low rate of return on environmental investment. The cost recovery problem means that environmental loans and grants need to be buttressed with innovative financial mechanisms such as the World Bank's revolving environmental fund for industrial enterprises in Liaoning. If the problems relating to supervisory control could be addressed, this would provide a useful model for other industrial cities across China. Some kind of micro-financing initiatives are also essential for farmers and small communities. It is unrealistic to assume that new eco-agricultural practices could be introduced without sufficient long-term financial support.

Second, *local awareness of the actual benefits of environmental capacity building has a dramatic impact on its effectiveness*. Donors tend to be more focused upon tangible rather than intangible project outcomes. However, changing attitudes and perceptions on the part of the beneficiaries involved in donor-funded environmental projects is an important measure of their success. This was clearly illustrated in the UNDP's environmental projects in Shenyang and Benxi. Unless new ideas and practices are embedded within the local consciousness they are unlikely to endure over time. Hence, incorporating attitudinal change into project assessment methods and evaluation is critical.

Third, *environmental capacity building is ill-suited to the existing structures and administrative procedures within donor institutions*. It requires sustained long-term commitment which cannot be realized on the basis of administrative procedures that are essentially geared towards short-term results and easily measurable outcomes. For environmental capacity building to be effective, donors need to build capacity requirements into their project feasibility studies and post-evaluations. More attention should be given to the sectoral allocation of environmental funding, especially in relation to the urban/rural divide. Environmental loans and grants need to be coordinated, and career incentives should be introduced to encourage staff to foster successful project outcomes. Above all, inter-donor coordination efforts need to be targeted towards capacity building.

Fourth, *developing environmental capacity in China is more likely to be effective when a certain level of institutional capacity already exists.* To many, this would appear to be a tautology. Yet, it is important to acknowledge that capacity building is multi-faceted – strengthening capacity in one area such as interagency coordination can often offset a weakness in another area such as knowledge acquisition. Moreover, capacity building, by definition, is a socially interactive process that relies upon 'learning by doing' rather than the replication of one particular model.

The real paradox for international donors is this: solving environmental problems in China requires capacity building, but in order for this to succeed, pre-existing capacity is necessary which cautions against targeting China's poorest regions and, therefore, runs counter to the overriding development goal of poverty alleviation. In poorer regions, the hope lies in a long-term approach to capacity building that relies upon consensus building, human development skills and training, innovative financial mechanisms and cost-effective technologies. Investment in environmental infrastructure is also necessary to create a political incentive for local leaders to become involved.

At a more micro level, the participatory ownership approach (currently being tested by the World Bank) is likely to yield some positive results in county and township settings. But this also cannot escape the inherent contradiction involved in facilitating institutional capacities via an external agency. The distinction between facilitating and controlling capacity building is a fine one. It involves an intricate dance between persuasion on the one hand and genuine participation on the other: too much donor intervention can undermine local autonomy and thereby have negative consequences for local commitment and participation, whereas too little runs the risk of 'local capture' by powerful interest groups and the diversion of investment funds. For capacity building to succeed, therefore, a great deal of institutional flexibility is required both on the part of the donors and the beneficiaries.

Clearly developing environmental capacity, especially in China's poorer regions, presents an enormous challenge to international donors, the Chinese government and its citizens. But to conclude on a more positive note, international environmental assistance is still relatively new and the positive outcomes identified in this study, no matter how limited, are an encouraging sign. At least in the world's most populous nation, where environmental problems are verging on the catastrophic, it may have a strong cumulative effect. To date, international environmental assistance to China has had far reaching implications insofar as local attitudes to environmental management are concerned. In this sense, donor efforts are a harbinger of things to come: a deeper recognition within China that managing environmental problems requires more than effective government intervention; it also requires strong economic incentives, participatory practices, and a sense of shared responsibility on the part of the stakeholders involved.

Epilogue
Can lessons be learned?

Scholars working on contemporary China enjoy the benefits of living in interesting times. China's ongoing transition to a socialist market economy has affected social, political and economic life in ways that we are only beginning to fully appreciate. Every visit to China yields new and interesting developments. But keeping pace with change across diverse regions of China also presents an enormous challenge. Our theories and not only our facts require regular revision. In writing this book, I have been struck by both how much and how little things have changed.

The question of whether China should take care of its environment in the immediate term rather than wait until economic development goals have been fully achieved is no longer a viable option. Unchecked environmental pollution, the finite nature of China's resource base, and growing urbanization mean that major adjustments have to be made along the way. The urban share of the population is expected to rise to 45 per cent by 2010. How will Chinese cities meet their rising water and energy needs as well as the expanding demand for environmental services? Improving urban environmental management will be critical but not sufficient. Over the coming years, equal attention will need to be paid to improving rural environmental management. Basic access to water and sanitation is still severely limited – in rural China approximately 360 million people are without access to clean water.[1] If this problem is not addressed, it is likely to encourage a further surge of environmental migrants to the overcrowded cities. It is already clear that the ecologically stressed western regions of China are no longer able to support their populations.[2]

China's new development concept, *xiaokang shehui* (all-round well off society), announced in 2002, calls for harmonious development between man and nature. Integrating environmental concerns into development planning and implementation is now state policy. The critical question is how can this best be achieved? This study has shown that a more decentralized and pluralistic approach to dealing with environmental problems is essential. Over the past five years, the shift away from a purely state-centric approach has become more easily recognizable. Chinese enterprises are increasingly keen to show off their green credentials. In November 2004, SEPA awarded environmentally friendly status to eight leading industries.[3] This new trend is in part being driven by international corporations working in China such as Dow Chemical that recently sponsored a programme worth US$720,000 to implement cleaner production amongst a number of medium-sized enterprises.[4]

Environmental NGOs have also risen to greater prominence. Their collective potential became apparent during a campaign in 2003 to halt a controversial plan to build 13 hydroelectric dams on the Nujiang, a UNESCO world heritage site, in Yunnan province. This river basin is home to a variety of endangered plants and species as well as 22 ethnic minorities. The coordinated campaign to mobilize public support led to a central government decision in February 2004 to place a moratorium on the plan until further assessments had been carried out.[5] At the time, this was seen as a landmark victory for China's environmental movement; it may well turn out to be a watershed in the history of NGO development in China.

However, a shift in the direction of a more pluralistic approach to environmental management is less discernible in the poorer, more ecologically fragile, and insecure regions of western China. To date, the *xibu da kaifa* (great opening of the West) campaign has focused almost exclusively on infrastructure development with limited attention given to environmental protection let alone capacity building. Pioneering efforts are being made by a few corporations to promote environmentally friendly investment, and a rising number of grassroots NGOs scattered across the provinces of Qinghai, Sichuan and Yunnan also offer a beacon of hope. But in the absence of reform from above, the question remains as to whether such local efforts will be enough to bring about a genuine change in both development policy and practice.

The re-direction of international environmental aid to western China offers an opportunity to advocate for reform. This imperative has become stronger in light of the fact that international aid to China is on the wane. In setting out to do the research for this book, I took the position that developing countries could not be reasonably expected to comply with international environmental standards without international support to help strengthen their domestic capacity for dealing with environmental problems. However, the issue has become more complicated in the case of China. The Chinese economy has grown to become the seventh largest in the world. To outsiders it is far more likely to be seen as a burgeoning economic powerhouse than a developing country desperately in need of international aid transfers. The successful Shenzou 5 space launch in October 2003 helped to reinforce this perception. At the time, the *Economist* was quick to question whether international aid to China could still be justified in arguing that China 'may still be a poor, if fast-growing economy, but if it chooses to spend its money on space travel there can be no good reason for outsiders to subsidise that choice'.[6]

Uneven development leading to growing disparities between the east and west and urban and rural areas of China alongside the ongoing environmental crisis still provide sufficient grounds for donor intervention. Nevertheless, it is becoming clear that international aid to China is on the decline. The World Food Program has already announced its withdrawal of food aid; and in March 2005 the Japanese Foreign Minister, Machimura Nobutaka, announced that Japanese yen loans to China would cease in fiscal 2008 to coincide with the hosting of the

Olympic Games.[7] During the remaining years, the environment is likely to remain a high priority for donors. They will inevitably be concerned with increasing the effectiveness of their environmental aid and its influence on policy-making in Beijing. For this to happen, lessons will need to be generated from below as well as from above.

But in contrast to the changes taking place at the policy level, the actual implementation of environmental aid remains slow and incremental. As of early 2005, many of the projects that I investigated for the purposes of this book had still not been completed. Although I was able to obtain evaluation reports for the UNDP projects (incorporated into the case study), these were still not available for the Japanese projects and only one of the World Bank projects had been evaluated.[8] The evaluation report for the World Bank's environmental improvement project in Liaoning is mainly concerned with the physical infrastructure component. With respect to institutional reform, it reinforces the concern raised in this study over the financial viability of the Liaoning Environment Fund. It also highlights the need 'to take account of the readiness and capability of the targeted recipients to absorb and utilise new skills and technology'.[9]

The disparity between rapid change at the policy level and slow project implementation means that important lessons can easily be discarded. By the time evaluations have been completed, attention has been diverted to other concerns, and indeed other countries. If international aid is withdrawing from China what is the value of a study on environmental aid implementation? There is, of course, still time for adjustments to be made to ongoing projects. And, of course, the lessons may be useful for donors working in other developing and post-socialist states. But the real value lies in providing lessons to government agencies, corporations and NGOs (both local and international) working to protect the environment in China for the longer term.

We must also consider the fact that China is now becoming a significant donor as well as a recipient. To date, Chinese civilian aid to developing countries has been linked primarily to conventional infrastructure building, agriculture and food processing, but it is likely that this may change in the near future. For example, in 2003, Premier Wen Jiabao pledged to increase aid to Africa for the prevention of infectious diseases and natural disasters and for environmental protection.[10] Over the past decade, what has China gained from its experience as a large recipient of environmental aid that will help it to contribute to solving environmental problems both within and beyond its borders? If lessons are to be learned a feedback loop is essential. For this we cannot rely on any single donor agency evaluation report. Nor can we assume that such a feedback loop will become a reality without serious attention to the broader issues of transparency and accountability. Negative as well as positive project outcomes need to be fed back up to central decision-makers in Beijing. After all, our faith in arguing against environmental catastrophe lies in the importance we attach to social as well as technological learning. It is my hope that this book, in a modest way, will contribute to that learning process.

Glossary of Chinese terms

Transcription of Chinese characters into the Roman alphabet follows the pinyin system. Many environmental management terms such as stakeholder participation (*huanbao canyuzhi*) are difficult to translate accurately into Chinese. Sustainable development (*ke chixu fazhan*) literally means endurance. It is usually explained simply as the need for a grandfather to provide food and clothing for his grandson and in so doing, conserve his resources for the future. In general, *huanjing baohu*, or more colloquially *huanbao*, refers to environmental protection and it is often used to refer to a broad-based definition of environmental management. Certain terms in Chinese are also difficult to translate into English. Some terms such as *danwei* (work unit), *hukou* (household registration) or *xia gang* (laid off workers) are best rendered in their Chinese original.

Banguanfangzuzhi	Government owned NGOs (GONGOs)
Changjiang	Yangzte River
Chengshi	City
Difang	Local
Gongmin shehui	Civil society
Huang Long	Yellow Dragon
Huang he	Yellow River
Huanjing baohu	Environmental protection
Huanjing guanli	Environmental management
Huanjing yishi	Environmental awareness
Hukou	Household registration
Jijinhui	Foundations
Jumin weiyuanhui	Urban residents committee
Ke chixu fazhan	Sustainable development
Liudong renkou	Floating population
Minban fei qiye danwei	Private non-profit organizations
Minjian zuzhi	Non-governmental organizations (NGOs)
Nanshui beidiao	South–north water transfer scheme
Nengli jianshe	Capacity building
Qunzhong yishi	Mass consciousness

Qunzhong zuzhi	Mass-based organizations
Renmin Daxue	People's University
Renmin Ribao	People's Daily
San Xia	Three Gorges Dam
Shehui tuanti	Social organizations
Weixing chengzhen	Satellite town
Xian	County
Xian zhen	County town
Xiangzhen qiye	Township (non-farmer) enterprise
Zhongyang	Centre
Zhengfu zhudao, gongtong canyu	Joint participation under the guidance of the government

Notes

Introduction

1 Qu Geping, 'China's environmental policy and world environmental problems', *International Environmental Affairs*, 2(2), 1990, p. 108.

2 Acid rain is formed when SO_2 combines with nitrogen oxide (NO_x) and becomes oxidized in the atmosphere to form sulphuric and nitric acids, which are then deposited on the earth's surface.

3 For surveys of China's environmental problems see Vaclav Smil, *The Bad Earth: Environmental Degradation in China*, Armonk, NY: M.E. Sharpe, 1984; He Bochan, *China on the Edge: Crisis of Ecology and Development*, San Francisco, CA: China Books and Periodicals, 1991; Vaclav Smil, *China's Environmental Crisis: An Inquiry into the Limits of National Development*, Armonk, NY: M.E. Sharpe, 1993; Richard Louis Edmonds, *Patterns of China's Lost Harmony: A Survey of the Country's Environmental Degradation and Protection*, London: Routledge, 1994; Qu Geping and Li Jinchang, *Population and Environment in China*, Boulder, CO: Lynne Rienner, 1994; Richard Louis Edmonds (ed.), *Managing the Chinese Environment*, New York: Oxford University Press, 1998. More recent works include Elizabeth C. Economy, *The River Runs Black: The Environmental Challenge to China's Future*, Ithaca, NY and London: Cornell University Press, 2004; and Vaclav Smil, *China's Past, China's Future: Energy, Food, Environment*, New York: Routledge, 2003.

4 On a 'business as usual basis', this contribution is expected to rise to 17 per cent in 2050 and 28 per cent in 2100. See Z.X. Zhang and Henk Folmer, 'The Chinese energy system: implications for future carbon dioxide emissions in China', *Journal of Energy and Development*, 21(1), 1996. However, recent reductions in CO_2 emissions in China cast considerable doubt upon the extent to which China's global contribution is likely to rise. Between 1992 and 2002, according to a report published in 2001 by the National Resources Defense Council in New York, aggregate CO_2 emissions in China grew by 8.4 per cent compared with 14 per cent in the United States. More significantly, during the same period, China's economy grew four times faster than that of the United States. See National Resources Defense Council, 'Second analysis confirms greenhouse gas reductions in China', www.nrdc.org/globalwarming/achinagg.asp (accessed 5 April 2002).

5 This book focuses upon development assistance that includes official development assistance (grants and concessionary loans) and development finance or non-concessionary loans such as those provided by the World Bank. According to the Organisation for Economic Cooperation and Development (OECD) rules, in order to qualify as official development assistance the grant element of the loan must be at least 25 per cent of the total loan value. The terms development assistance and development aid will be used interchangeably throughout the book.

6 UNDP China, *China Environment and Sustainable Development Resource Book II: A Compendium of Donor Activities*, Beijing: UNDP, 1996. Comprehensive figures for environmental lending after 1996 are currently unavailable. Environmental projects (broadly defined) first started in China in the late 1980s and focused largely upon water supply issues. By 1992, donors were giving greater priority to air pollution and environmental infrastructure-related projects and this continued throughout the 1990s.

7 World Bank, China, *The World Bank Group in China, Facts and Figures*, Beijing: World Bank, 1999.

8 Calculated on the basis of Japan's environmental loan projects to China listed in JBIC, *JBIC Annual Report*, Tokyo: JBIC, 2002.

9 JBIC, 'Supporting environmental conservation and human resource development in China – FY2003 ODA loan package for China', www.jbic.go.jp/autocontents/english/news/2004/000025/index.htm (accessed 5 October 2004).

10 The few studies that are available have focused upon Central and Eastern Europe, see Martin Jänicke and Hans Weidner (eds), *National Environmental Policies: A Comparative Study of Capacity-Building*, Berlin: Springer, 1997; Susan Baker and Petr Jehlicka, 'Dilemmas of transition: the environment, democracy and economic reform in east central Europe – an introduction', *Environmental Politics*, 7(1), 1998; and Elizabeth Wilson, 'Capacity for environmental action in Slovakia', *Journal of Environmental Planning and Management*, 42(4), 1999.

11 See Andrew Hurrell and Benedict Kingsbury (eds), *The International Politics of the Environment: Actors, Interests and Institutions*, Oxford: Clarendon Press, 1992; Ronnie Lipschutz and Ken Conca (eds), *The State and Social Power in Global Environmental Politics*, New York: Columbia University Press, 1993; Andrew Hurrell, 'A crisis of ecological viability? Global environmental change and the nation state', in John Dunn (ed.), *Contemporary Crisis of the Nation State?*, Oxford: Blackwell Publishers, 1995; and Richard Sandbrook, 'UNGASS has run out of steam', *International Affairs*, 73(4), 1997.

12 By compliance I mean the voluntary observance of rules rather than an attempt to enforce rules on the basis of coercion. In the case of the environment, a coercive approach is highly likely to be counterproductive because of the overriding need to nurture public support.

13 The few exceptions have centred upon Japanese environmental assistance to China with an emphasis upon levels of assistance and policies rather than implementation. They include Peter Evans, 'Japan's green aid', *China Business Review*, July–August, 1994; Matsuura Shigenori, 'China's air pollution and Japan's response to it', *International Environmental Affairs*, 7(3), 1995; and Susan Pharr and Ming Wan, 'Yen for the earth: Japan's pro-active China environmental policy', in Michael B. McElroy, Chris P. Nielsen and Peter Lydon (eds), *Energizing China: Reconciling Environmental Protection and Economic Growth*, Cambridge, MA: Harvard University Press, 1998.

14 For studies focusing on regulatory enforcement see Richard Lotspeich and Aimin Chen, 'Environmental protection in the People's Republic of China', *Journal of Contemporary China*, 6(14), 1997; and Zhang Weijong, Ilan Vertinsky, Terry Ursacki and Peter Newetz, 'Can China be a clean tiger?: Growth strategies and environmental realities', *Pacific Affairs*, 72(1), 1999. Those studies concerned with industry compliance and the impact of economic decentralization include Xiaoying Ma and Leonard Ortolano, *Environmental Regulation in China: Institutions, Enforcement, and Compliance*, Lanham, MD: Rowman and Littlefield, 2000; Scott Rozelle, Xiaoyang Ma and Leonard Ortolano, 'Industrial wastewater control in Chinese cities: determinants of success in environmental policy', *Natural Resources Modeling*, 7(4), 1993; and Abigail R. Jahiel, 'The contradictory impact of reform on environmental protection in China', *China Quarterly*, 149(March), 1997.

15 See Hon Chan, K.C. Cheung and Jack M.K. Lo, 'Environmental control in the PRC', in Stuart Nagel and Miriam Mills (eds), *Public Policy in China*, Westport, CT: Greenwood Press, 1993; Barbara Sinkule and Leonard Ortolano, *Implementing Environmental Policy in China*, Westport, CT: Praeger, 1995; Shui-Yan Tang, Carles Wing-Hung Lo, Kai-Chee Lo Cheung and Jack Man-Keung, 'Institutional constraints on environmental management in urban China: environmental impact assessment in Guangzhou and Shanghai', *China Quarterly*, 152(December), 1997; and Abigail R. Jahiel, 'The organization of environmental protection in China', *China Quarterly*, 156(December), 1998.

16 In the development literature 'local' generally refers to a social entity (i.e. community, village or groups with strong kinship connections and interpersonal relationships). Here 'local' refers to a political entity (i.e. groups at the sub-national level where social relationships are more formally constructed). In the Chinese language a distinction is made between *zhongyang* (centre) and *difang* (local). *Zhongyang* includes the State Council, Party Central Committee and Commissions and *difang* includes provinces, municipalities, counties and townships.

17 Japan did receive some funds from the United States – a total of US$2 billion in economic aid largely for foodstuffs and materials. See John Dower, *Embracing Defeat: Japan in the Aftermath of World War II*, London: Allen Lane, 1999, p. 529.

18 I am grateful to Professor Masayoshi Matsumura from Ritsumeikan University in Japan for drawing my attention to this point.

19 The fieldwork involved visits to international aid agencies in Tokyo, New York and Washington, DC, in September 1997 and March 1999 as well as two separate visits to China between May and July 1998, and April and July 1999. I conducted over 140 interviews in Beijing and 11 other cities across China in addition to visiting a number of counties and townships. At the local level, interviewees were mostly government officials from environmental, planning and construction agencies, in addition to factory managers, farmers and local residents.

20 The literature on development assistance failure is vast and largely empirical in content. Only a few studies will be mentioned here. For literature on poor donor administration and maintenance of development projects see Norman Uphoff, *Local Institutional Development*, West Hartford, CT: Kumarian Press, 1986; David Korten, *Getting to the 21st Century: Voluntary Action and the Global Agenda*, West Hartford, CT: Kumarian Press, 1990; and Robert Cassen, *Does Aid Work?*, 2nd edn, Oxford: Clarendon Press, 1994. Robert Wade provides a convincing account of the strategic behaviour of the recipient administration on the basis of empirical studies in India. See Robert Wade, 'The market for public office: why the Indian state is not better at development', *World Development*, 13(4), 1985.

21 See Michio Hashimoto, 'Development and environmental problems', *Asian Economic Journal*, 8(1), 1994; and David Potter, 'Assessing Japan's environmental aid policy', *Pacific Affairs*, 67(2), 1994.

22 Jun Nishikawa, 'Reform of foreign aid', *Nikkei Weekly*, 12 May 1997.

23 See Korten, *Getting to the 21st Century*; Bruce M. Rich, *Mortgaging the Earth: The World Bank, Environmental Impoverishment, and the Crisis of Development*, Boston, MA: Beacon Press, 1994; and Catherine Caufield, *Masters of Illusion: The World Bank and the Poverty of the Nations*, London: Pan Books, 1998.

1 Developing environmental capacity

1 Mancur Olson, *The Logic of Collective Action*, Cambridge, MA: Harvard University Press, 1965.

2 In Hardin's seminal work on the 'tragedy of the commons' each herder seeks to maximise his returns from grazing cattle on common land even beyond the point at which the carrying capacity of the land is reached, therefore leading to 'remorseless

tragedy for all'. Note here that Hardin was concerned not with environmental protection but with population growth and that his solutions were 'mutual coercion, mutually agreed upon' and 'restraints on the freedom to breed', as he put it. See Garrett Hardin, 'The tragedy of the commons', *Science*, 162(3859), 1968.

3 See William Ophuls, 'Leviathan or oblivion', in Herman Daly (ed.), *Toward a Steady-State Economy*, San Francisco, CA: Freeman, 1973; and Ian Carruthers and Roy Stoner, 'Economic aspects and policy issues in ground water development', World Bank Staff Working Paper No. 496, Washington, DC: World Bank, 1981.

4 See Richard Falk, 'The global promise of social movements: explorations at the edge of time', *Alternatives*, 12(2), 1987; John Clarke, *Democratizing Development: The Role of Voluntary Organization*, London: Earthscan, 1991; and Paul Ekins, *A New World Order: Grassroots Movements for Global Change*, London: Routledge, 1992.

5 In particular, the feedback between expected resource scarcity and efforts towards technological innovation. Heinz Arndt makes the point that feedback not only operates through the price mechanism but also through socio-political responses such as the environmentalist movement. Heinz W. Arndt, 'Sustainable development and the discount rate', in Heinz W. Arndt, *50 Years of Development Studies*, Canberra: National Centre for Development Studies, Australian National University, 1993.

6 World Commission on Environment and Development, *Our Common Future*, Oxford: Oxford University Press, 1987.

7 See David Pearce, Anil Markandya and Edward Barbier, *Blueprint for a Green Economy*, London: Earthscan, 1989; Tom H. Tietenberg, 'Economic instruments for environmental regulation', *Oxford Review of Economic Policy*, 6(1), 1990; and Theodore Panayotou, *Instruments of Change: Motivating and Financing Sustainable Development*, London: Earthscan, 1998. It is important to note here that economists are by no means united in their response to solving environmental problems. The property rights school or 'free market environmentalism' and 'ecological economics' offer two alternative positions. The former advocate competitive markets based upon efficient price signals and transferable property rights (see Terry Anderson and Donald Leal, *Free Market Environmentalism*, Boulder, CO: Westview, 1991; and Alan Moran, Andrew Chisholm and Michael Porter (eds), *Markets, Resources and the Environment*, North Sydney: Allen and Unwin, 1991) while the latter also stress the finite nature of the earth's resource base (see Nicholas Georgescu-Roegan, *The Entropy Law and the Economic Process*, Cambridge, MA: Harvard University Press, 1971; Herman Daly (ed.), *Steady-State Economics*, San Francisco, CA: W.H. Freeman, 1977; and Kenneth Arrow, Bert Bolin, Robert Costanza, Partha Dasgupta, Carl Folke, C.S. Holling, Bengt-Owe Jansson, Simon Levin, Karl-Goran Maler, Charles Perrings and David Pimental, 'Economic growth, carrying capacity, and the environment', *Science*, 268(5210), 1995).

8 Arthur Pigou, *The Economics of Welfare*, London: Macmillan, 1932, first published 1920.

9 This does not suggest that environmental economists are not concerned with valuing the environment. However, given the lack of agreement over methodologies, in addition to continuing scientific uncertainty over environmental impacts, economic instruments do not depend upon a monetary value but instead rely upon government standards.

10 Tom H. Tietenberg, *Emissions Trading: An Exercise in Reforming Pollution Policy*, Washington, DC: Resources for the Future, 1985.

11 As Robyn Eckersley notes, 'taxes are merely technical instruments based on scientific/political compromises, not economic theory'. Robyn Eckersley (ed.), *Markets, the State and the Environment: Towards Integration*, South Melbourne: Macmillan Education Australia, 1995, p. 14.

12 Michael Redclift, 'Sustainable development and popular participation: a framework for analysis', in Dharam Ghai and Jessica Vivian (eds), *Grassroots Environmental Action: People's Participation in Sustainable Development*, London: Routledge, 1992, p. 29.

13 Theodore Panayotou, 'Economic incentives for environmental management in developing countries', in OECD, *Economic Instruments for Environmental Management in Developing Countries*, Paris: OECD, 1992.

14 Robert Goodin, 'Selling environmental indulgences', *Kyklos*, 47(4), 1994.

15 Michael Jacobs, 'Sustainability and "the market": a typology of environmental economics', in Eckersley (ed.), *Markets, the State and the Environment*, p. 68.

16 Mark Sagoff, *The Economy of the Earth: Philosophy, Law, and the Environment*, Cambridge: Cambridge University Press, 1988, p. 28. Sagoff's interpretation echoes Alexis de Tocqueville's proposition that self-interest and altruism coexist, or the notion of 'self-interest properly understood'. In other words, individuals are by nature self-interested but also obligated to act as citizens and participate in their own governance. See Eduardo Nolla, *De la democratie en Amerique*, Paris: J. Vrin, 1990. Equally, the notion of dualism is central to the work of Emile Durkheim and his treatise on the need to counterbalance impersonal constraints, 'contractual solidarity' with personal 'organic solidarity'. See Emile Durkheim, [On the] *Division of Labour in Society*, trans. G. Simpson, New York: Macmillan, 1933, first published 1911.

17 See Barry Commoner, *The Closing Circle: Confronting the Environmental Crisis*, London: Cape, 1972.

18 See Adolf Gunderson, *The Environmental Promise of Democratic Deliberation*, Madison, WI: University of Wisconsin Press, 1995; Amy Douglas, *The Politics of Environmental Mediation*, New York: Columbia University Press, 1987; and John Dryzek, *The Politics of the Earth: Environmental Discourses*, Oxford: Oxford University Press, 1997.

19 There exists a contentious debate within the participation literature over whether participation should be perceived as *instrumental* or *intrinsic*. It is fair to say that in general those advocates of the instrumental perspective are not true believers in the democratic approach. In the Civic Republican tradition participation is intrinsic. J.S. Mill favoured the intrinsic perspective, as he believed that it provided a counter-balance to 'passivity, inertia, timidity and intellectual stagnation'. In a similar vein Albert Hirschman's view on participation was that one 'can raise the benefit accruing to him…by stepping up *his own input*' (both cited in Albert O. Hirschman, *Shifting Involvements: Private Interest and Public Action*, Oxford: Oxford University Press, 1982, pp. 90 and 86 respectively, emphasis in original). But whether instrumental or intrinsic, what matters is that participation in practice takes place on a voluntary basis.

20 Dryzek, *The Politics of the Earth*.

21 Martin Jänicke, 'Conditions for environmental policy success: an international comparison', *The Environmentalist*, 12(1), 1992.

22 Charles Ziegler (ed.), *Environmental Policy in the USSR*, Amherst, MA: University of Massachusetts, 1987.

23 See Rajni Kothari, *Rethinking Development: In Search of Humane Alternatives*, New York: New Horizons Press, 1989; Korten, *Getting to the 21st Century*; and Dharam Ghai and Jessica Vivian (eds), *Grassroots Environmental Action: People's Participation in Sustainable Development*, London: Routledge, 1992.

24 The epistemological dimension of community participation is an important focus in the development literature. See Appel Marglin and Stephen Marglin (eds), *Dominating Knowledge: Development, Culture and Resistance*, Oxford: Clarendon Press, 1990. However, local knowledge does have its limitations. For an interesting critique of the indigenous versus science dichotomy see Arun Agrawal, 'Dismantling the divide between indigenous and scientific knowledge', *Development and Change*, 26(3), 1995.

25 Lore Ruttan, 'Closing the commons: cooperation for gain or restraint?', *Human Ecology*, 26(1), 1998, p. 51.

26 Andrew Shepherd, 'Participatory environmental management: contradiction of process, project and bureaucracy in the Himalayan foothills', *Public Administration and Development*, 15(4), 1995.

27 The theory of ecological modernization is derived from the comparative literature on environmental policy-making with its origins in the works of Joseph Huber, *Die Verlorene Unchuld der Okologie*, Frankfurt: Fischer Verlag, 1982; and Martin Jänicke, *Preventative Environmental Policy as Ecological Modernisation and Structural Policy*, Berlin: Berlin Science Center, 1985.

28 See Gert Spaargaren and Arthur Mol, 'Sociology, environment, and modernity: ecological modernization as a theory of social change', *Society and Natural Resources*, 5(4), 1992; Albert Weale, *The New Politics of Pollution*, Manchester: Manchester University Press, 1992; and Maarten Hajer, *The Politics of Environmental Discourse: Ecological Modernization and the Policy Process*, Oxford: Clarendon Press, 1995.

29 Andrew Gouldson and Joseph Murphy, *Regulatory Realities: The Implementation and Impact of Industrial Environmental Regulation*, London: Earthscan, 1998.

30 Ibid., p. 82.

31 Hajer, *The Politics of Environmental Discourse*.

32 See Ulrich Beck, *The Risk Society: Towards a New Modernity*, trans. Mark Ritter, London: Sage, 1992; and Ulrich Beck, Anthony Giddens and Scott Lash, *Reflexive Modernization: Politics, Tradition and Aesthetics in the Modern Social Order*, Cambridge: Polity Press in association with Blackwell Publishers, 1994.

33 Beck, *The Risk Society*, p. 22.

34 Beck *et al.*, *Reflexive Modernization*.

35 Arthur Mol, 'Ecological modernisation and institutional reflexivity: environmental reform in the late modern age', *Environmental Politics*, 5(2), 1996.

36 The literature on institution building is large. Pioneering works include: Milton Esman, 'Institution building in national development', in Gove Hambridge (ed.), *Dynamics of Development*, New York: Praeger, 1964; George F. Gant, 'The institution building project', *International Review of Administrative Sciences*, 32(3), 1966; Arnold Rivkin (ed.), *Nations by Design: Institution Building in Africa*, Garden City, NY: Doubleday, 1968; and Joseph Eaton (ed.), *Institution Building and Development: From Concepts to Application*, Beverly Hills, CA: Sage Publications, 1972.

37 See Lucian Pye, *Aspects of Political Development: An Analytic Study*, Boston, MA: Little Brown, 1966; Samuel Huntington, *Political Order in Changing Societies*, New Haven, CT: Yale University Press, 1968; and Leonard Binder, James S. Coleman, Joseph LaPalombara and Lucian W. Pye, *Crises and Sequences in Political Development*, Princeton: Princeton University Press, 1971.

38 See Milton Esman, *Administration and Development in Malaysia: Institution Building and Reform in a Plural Society*, Ithaca, NY: Cornell University Press, 1972.

39 David Fairman and Michael Ross, 'Old fads, new lessons: learning from economic development assistance', in Robert Keohane and Marc Levy (eds), *Institutions for Environmental Aid: Pitfalls and Promise*, Cambridge, MA: MIT Press, 1996, p. 4.

40 See Robert Cassen, *Does Aid Work?*, 2nd edn, Oxford: Clarendon Press, 1994; Arturo Israel, *Institutional Development: Incentives to Performance*, Baltimore, MD: Johns Hopkins University Press, 1988; Milton Esman, *Management Dimensions of Development: Perspectives and Strategies*, West Hartford, CT: Kumarian Press, 1991.

41 John Cohen, 'Foreign advisors and capacity building: the case of Kenya', *Public Administration and Development*, 12(5), 1993.

42 I am cognizant of the fact that many recipient countries are highly experienced in local environmental resource management. It is, therefore, important to distinguish between local natural resource problems and other environmental problems resulting from rapid economic growth or global environmental change where experience or expertise is often lacking.

43 Deborah Eade, *Capacity-Building: An Approach to People-Centred Development*, Oxford: Oxfam (UK and Ireland), 1997, p. 2.

44 Jay M. Shafritz, *Dictionary of Public Administration*, New York: Oxford University Press, 1986, p. 79.

45 Merilee Grindle and Mary Hildebrand, 'Building sustainable capacity in the public sector: what can be done?', *Public Administration and Development*, 15(5), 1995.

46 This way of thinking has been inspired by the work of Amartya Sen who distinguishes between the ends and means of the role of freedom in development. See Amartya Sen, 'The ends and means of development', in Amartya Sen (ed.), *Development as Freedom*, Oxford: Oxford University Press, 1999.

47 Amartya Sen, 'Public action and the quality of life in developing countries', *Oxford Bulletin of Economics and Statistics*, 43(4), 1981.

48 OECD, DAC, *Development Cooperation: Efforts and Policies of the Members of Development Assistance Committee*, DAC Report, Paris: OECD, 1996.

49 The World Bank has used the term environmental capacity building to refer to institutional development for environmental purposes. See Sergio Margulis and Tonje Vetleseter, 'Environmental capacity building – a portfolio review', *Environment Matters at the World Bank: Annual Review*, Washington, DC: World Bank, Fall 1998. And the OECD has used the term 'capacity for environmental protection' which it defines very broadly as 'a society's ability to identify and solve environmental problems'. OECD, DAC, *1994 Environmental Policy Review*, DAC Report, Paris: OECD, 1995, p. 8.

50 The notion of capacity building is generally associated with developing countries. Yet, as noted by Martin Jänicke and Hans Weidner (eds), *National Environmental Policies: A Comparative Study of Capacity-Building*, Berlin: Springer, 1997, p. 2, the concept is equally applicable to the developed world. Unfortunately, very little has been written on the subject of environmental capacity amongst developed countries upon which to draw any relevant lessons.

51 Paul Mosley, Jane Harrigan and John Toye, *Aid and Power: The World Bank and Policy-Based Lending*, volumes 1 and 2, London: Routledge, 1991.

52 See Alex Duncan, 'Aid effectiveness in raising adaptive capacity in the low-income countries', in John Lewis and Valeriana Kallab (eds), *Development Strategies Reconsidered*, Washington, DC: Overseas Development Council, 1985; World Bank, *World Bank Development Report: Development and the Environment*, Washington, DC: World Bank, 1992; and Klye Danish, 'The promise of national environmental funds in developing countries', *International Environmental Affairs*, 7(2), 1995.

53 Ronnie Lipschutz and Ken Conca (eds), *The State and Social Power in Global Environmental Politics*, New York: Columbia University Press, 1993.

54 The UNDP's Capacity 21 Fund was launched at the United Nations Conference on Environment and Development in 1992 with the purpose of assisting developing countries with integrating the principles of sustainable development (Agenda 21) into their national policies and development programmes. The institutional emphasis is upon stakeholder participation and information sharing. See UNDP, *The Fifth Year: Learning and Growing*, Annual Report Capacity 21, New York: UNDP, 1998. For the World Bank, building institutional capacity for environmental management is equivalent to building state capacity per se through 'reinvigorating public institutions' which, in turn, are strengthened by 'providing incentives for public officials to perform better while keeping arbitrary action in check'. World Bank, *World Development Report 1997: The State in a Changing World*, Washington, DC: World Bank, 1997.

55 See Jänicke and Weidner (eds), *National Environmental Policies*; Susan Baker and Petr Jehlicka, 'Dilemmas of transition: the environment, democracy and economic reform in east central Europe – an introduction', *Environmental Politics*, 7(1), 1998; and Elizabeth Wilson, 'Capacity for environmental action in Slovakia', *Journal of Environmental Planning and Management*, 42(4), 1999.

56 For example, in the early 1990s financial assistance provided by the World Bank and the United Sates for logging reform in the Philippines was successful because the then Corazon Aquino-led government was committed to the reform from the outset. But in the case of Indonesia, the government refused to accept a forestry loan from the World Bank because it was reluctant to interfere with entrenched domestic logging interests. Peter M. Haas, Robert Keohane and Marc Levy, *Institutions for the Earth: Sources of Effective International Environmental Protection*, Cambridge, MA: MIT Press, 1993.

57 Fairman and Ross, 'Old fads, new lessons', p. 45.

58 José Goldenberg and Thomas Johansson, *Energy as an Instrument for Socio-Economic Development*, New York: UNDP, 1995; and Amulya Reddy, Robert Williams and Thomas Johansson, *Energy After Rio: Prospects and Challenges*, New York: UNDP, 1997.

59 Epistemic communities can be national or transnational. They are defined by Peter Haas as 'a network of professionals with recognised expertise and competence in a particular domain'. Peter M. Haas, 'Introduction: epistemic communities and international policy coordination', *International Organization*, 46(1), 1992, p. 3.

60 See Tom H. Tietenberg and David Wheeler, 'Empowering the community: information strategies for pollution control', paper presented at the Frontiers of Environmental Economics Conference, Airlie House, Virginia, 23–25 October 1998.

61 For example, studies carried out by the Economics of Industrial Pollution Control Research Team at the World Bank have revealed that public disclosure programmes in Indonesia and the Philippines are helping to facilitate environmental performance. The Program for Pollution Control, Evaluation, and Rating in Indonesia and Ecowatch in the Philippines are designed as public disclosure programmes to ally market forces with local communities in enforcing industrial pollution control. Factories are assigned a colour rating by government on the basis of their environmental performance (gold, blue and black). The benefits of the programmes are that local communities can better negotiate environmental arrangements, factories earn market rewards, investors and insurers can assess liabilities and regulators can concentrate limited resources upon the worst performers.

62 Jänicke and Weidner (eds), *National Environmental Policies*.

2 The long march towards environmental management in China

1 Abigail R. Jahiel, 'The contradictory impact of reform on environmental protection in China', *China Quarterly*, 149(March), 1997.

2 See William Kapp, *Environmental Policies and Development Planning in Contemporary China and Other Essays*, Paris: Mouton, 1974; and Norman Myers, 'China's approach to environmental conservation', *Environmental Affairs*, 5(33), 1976. A few years earlier a discordant note was struck by L. Orleans and Richard Suttmeier in a study that revealed SO_2 emissions in Beijing were leading to the enlargement of the liver amongst school children. See L. Orleans and Richard Suttmeier, 'The Mao ethic and environmental quality', *Science*, 170, 1970.

3 See Vaclav Smil, *The Bad Earth: Environmental Degradation in China*, Armonk, NY: M.E. Sharpe, 1984; Vaclav Smil, *China's Environmental Crisis: An Inquiry into the Limits of National Development*, Armonk, NY: M.E. Sharpe, 1993; Eduard B. Vermeer, 'Management of environmental pollution in China: problems and abatement measures', *China Information*, 5(1), 1990; He Bochan, *China on the Edge: Crisis of Ecology and Development*, San Francisco, CA: China Books and Periodicals, 1991; Richard Louis Edmonds, *Patterns of China's Lost Harmony: A Survey of the Country's Environmental Degradation and Protection*, London: Routledge, 1994; and Qu Geping and Li Jinchang, *Population and Environment in China*, Boulder, CO: Lynne Rienner, 1994. For a more recent survey see Elizabeth C. Economy, *The River Runs Black: The Environmental Challenge to China's Future*, Ithaca, NY and London: Cornell University Press, 2004.

4   The first nationwide survey of industrial pollution was not completed until 1988. See Qu Geping, 'China's industrial pollution survey', *China Reconstructs*, 37(8), 1988. The number of monitoring stations in China has increased two-fold during the reform period from 1,144 in 1984 to 2,223 in 1996. *Zhongguo huanjing nianjian 1996* (China Environmental Yearbook 1996), Beijing: Zhongguo huanjing chubanshe, 1996. SEPA now monitors emissions from 8 million firms but hundreds of thousands of industrial enterprises at the township and village level remain outside of their control. Recently installed Geographic Information Systems are helping national assessments by providing satellite imaging to reflect ground level problems.

5   Joseph Needham, *Science and Civilisation in China*, Cambridge: Cambridge University Press, 1965; and Joseph Needham, *Science in Traditional China: A Comparative Perspective*, Cambridge, MA: Harvard University Press, 1981. Over the past two millennia there have been at least 16,000 recorded floods in China – increasing in intensity since the Sui dynasty (AD 589–617). Edmonds, *Patterns of China's Lost Harmony*, p. 73.

6   Stanley Dennis Richardson, *Forests and Forestry in China: Changing Patterns of Resource Development*, Washington, DC: Island Press, 1990.

7   Mark Elvin and Liu Ts'ui-jung, *Sediments of Time: Environment and Society in Chinese History*, Cambridge: Cambridge University Press, 1998.

8   Mark Elvin, 'The environmental legacy of imperial China', *China Quarterly*, 156(December), 1998.

9   Shinkichi Eto, 'China and Sino-Japanese relations in the coming decades', *Japan Review of International Affairs*, 10(1), 1996.

10 Cited in Smil, *China's Environmental Crisis*, p. 14.

11 Nicholas Lardy, *Agricultural Prices in China's Modern Economic Development*, Cambridge: Cambridge University Press, 1983; Kam Wing Chan, *Cities with Invisible Walls: Reinterpreting Urbanization in Post-1949 China*, Hong Kong: Oxford University Press, 1994; and Gregory Eliyn Guldin (ed.), *Farewell to Peasant China: Rural Urbanization and Social Change in the Late Twentieth Century*, Armonk, NY: M.E. Sharpe, 1997.

12 According to a United Nations Children's Fund (UNICEF) annual report published in 1997, 400 million Chinese do not have access to safe drinking water and 76 per cent of the population lacks access to safe sanitation. China ranks alongside Cambodia, Nepal and Vietnam for providing toilets, and below India and Bangladesh in providing access to safe drinking water. Reuters, *South China Morning Post*, 24 July 1997.

13 For an excellent account of environmental degradation during the Mao era see Judith Shapiro, *Mao's War Against Nature: Politics and the Environment in Revolutionary China*, Cambridge: Cambridge University Press, 2001.

14 *Peking Review*, 27 June 1972, cited in Kapp, *Environmental Policies*, p. 52.

15 Kapp, *Environmental Policies*, p. 52.

16 For a good survey of the growth of TVEs see Christopher Findlay, Andrew Watson and Harry Wu (eds), *Rural Enterprises in China*, London: Macmillan, 1994.

17 Edmonds, *Patterns of China's Lost Harmony*.

18 See Xue Wei, 'Water resources and economic development in China', *Chinese Economic Studies*, 29(1), 1996. For a more recent update on China's water crisis see http://English.peopledaily.com.cn (accessed 6 June 2002).

19 The floods in the summers of 1994, 1995, 1996 and 1998 led to over 1,000 deaths per year. Droughts in China have been equally costly. A severe drought in northern China caused the *Huang* (Yellow) River to dry up on 19 occasions between 1970 and 1996 and for 333 days between 1990 and 1995. Direct economic losses to agriculture and industrial production were estimated to be around RMB20 billion (US$2.4 billion). Reuters, *Xinhua* Beijing, 23 July 1997.

20 Despite population control polices enacted in the post-reform period, population growth is still approximately 1.6 per cent (with a population increase of

15 million per annum). The population is projected to rise to 1.4 billion by 2010. Judith Banister contends that the impact of China's population upon its environment is not growth per se, which is reportedly stabilizing as a consequence of China's one child policy. She argues that an expanding 'economically active population' aged between 15 and 64 is increasing the need for greater employment and in turn the prospect of more environmentally damaging economic activity. See Judith Banister, 'Population, public health and the environment in China', *China Quarterly*, 156(December), 1998.

21 CITES only lists those species that are endangered by trade.

22 *Zhongguo Qingnian Bao* (China Youth Daily), 2 April 1999; and *Beijing* Review, 8–14 February 1999.

23 Total suspended particulates are airborne particulates caused by the combustion of fossil fuels and often combine with naturally occurring wind-blown dust as in the cities of Beijing and Huhhot. High levels of TSP can lead to chronic obstructive pulmonary disease.

24 World Bank, *Clear Water and Blue Skies: China's Environment in the Twenty-First Century*, Washington, DC: World Bank, 1997.

25 Michael B. McElroy, 'Industrial growth, air pollution, and environmental damage: complex challenges for China', in Michael B. McElroy, Chris P. Nielsen and Peter Lydon (eds), *Energizing China: Reconciling Environmental Protection and Economic Growth*, Cambridge, MA: Harvard University Press, 1998; and Japan, Ministry of Foreign Affairs, *Japan's Official Development Assistance Annual Report 1996*, Tokyo: Association for the Promotion of International Cooperation, 1997.

26 Dazhi Gao, Wang Jinjia and Cai Jinlui, 'Mercury pollution and control in China', *Journal of Environmental Sciences China*, (3), 1991.

27 World Bank, *China, Air, Land, and Water: Environmental Priorities for a New Millennium*, Washington, DC: World Bank, 2001.

28 The increasing share of municipal pollution reflects both a low level of environmental infrastructure (approximately 80 per cent of municipal wastewater and sewage is discharged untreated) and rising urbanization. The boundaries between urban and rural areas in China, however, are difficult to define. It is a complex issue because of the emergence of large numbers of migrant workers (referred to as China's floating population (*liudong renkou*)), the proliferation of satellite towns, and numerous township enterprises situated in rural areas. The usual method used to define urbanization in China is to count the number of non-agricultural households living in designated cities and towns. In 2000, this amounted to 36 per cent of the population. *China Daily*, 5 September 2002.

29 National Resources Defense Council, 'Second analysis confirms greenhouse gas reductions in China', www.nrdc.org/globalwarming/achinagg.asp (accessed 5 April 2002).

30 *Renmin Ribao*, 2 March 2002.

31 China reduced its reliance on coal from 75 per cent of total energy consumption in 1997 to 69 per cent in 2000. *Zhongguo tongji nianjian 2000* (China Statistical Yearbook 2000), Beijing: Zhongguo tongji chubanshe, 2000.

32 See McElroy, 'Industrial growth, air pollution'; and World Bank, *China, Air, Land, and Water*. It is worth noting here that the reduction of particulate matter from large industrial sources is relatively cost-effective compared to reducing emissions of sulphur compounds. In the former case electrostatic precipitators, wet scrubbers and pre-combustion cleaning are used, compared to the more expensive flue gas desulphurization or fluidized bed combustion used in the latter case. China's first low-pollution power station (with imported desulphurization equipment from Finland) began operation in July 1999 located in Jiangsu province. FBIS Beijing *Xinhua*, 21 July 1999.

33 *Huasheng Monthly*, 8 July 2002.

34 Michael B. McElroy, Chris P. Nielsen and Peter Lydon (eds), *Energizing China: Reconciling Environmental Protection and Economic Growth*, Cambridge, MA: Harvard University Press, 1998. The atmospheric chemistry of photochemical smog involves interactions between nitrogen oxides, hydrocarbons, carbon monoxide and sunlight.

35 Even small doses of lead poisoning can affect neurological development and physical growth in children. The study revealed a strong correlation between automobile emissions and elevated blood-lead levels. See Xiao-ming Shen, John F. Rosen, Di Guo Wu and Sheng-mei, 'Childhood lead poisoning in China', *The Science of the Total Environment*, 181(2), 1996. Note here that the Chinese government imposed a nationwide ban on leaded fuel on 1 January 2000.

36 Throughout this book, unless otherwise stated, official figures quoted in Renminbi have been converted to US dollars at the conversion rate of RMB8.27 to the dollar in 1999.

37 *Beijing Review*, 3–9 May 1999.

38 The World Bank uses a 'willingness to pay' methodology compared to the human capital methodology used by the Chinese. Unlike the latter, the 'willingness to pay' approach measures the value of human life and health in the market place. World Bank, *Clear Water and Blue Skies*, p. 24.

39 Vaclav Smil is careful to point out the difficulties involved in attempting to quantify the costs of environmental degradation in China including inconsistent accounting procedures, problematic valuations for lost services (e.g. wetlands as buffers against pollution), and the difficulty of quantifying the costs of human suffering. He also suggests that the 'willingness-to-pay' methodology adopted by the World Bank is clearly ill-suited to China's ongoing quest for personal wealth. Nevertheless, he stresses that such analytical exercises can provide useful data for policy formulation and assessment. Vaclav Smil, 'Environmental problems in China: estimates of economic costs', Special Report No. 5, Honolulu: East West Center, April 1996.

40 *Zhongguo Qingnian Bao* (China Youth Daily), 2 April 1999.

41 *Beijing Review*, 3–9 May 1999.

42 *Renmin Ribao*, 20 May 1999.

43 Cited in *Zhongguo huanjing nianjian 1995* (China Environmental Yearbook 1995), Beijing: Zhongguo huanjing chubanshe, 1995.

44 *Zhongguo huanjing nianjian 1996*, p. 3.

45 *China Daily*, 20 June 1991.

46 For example, in 1997 the China Communist Youth League (through its Development Foundation) launched the 'Green Hope Project' to fund large-scale afforestation programmes in the Huang He and Chang Jiang basins. FBIS Beijing *Xinhua*, 21 September 1999.

47 In this sense, China faces the same dilemma as the former socialist governments of Central and Eastern Europe. See Duncan Fisher, *Paradise Deferred: Environmental Policymaking in Central and Eastern Europe*, London: Energy and Environment Programme, Royal Institute of International Affairs, 1992.

48 *China Daily*, 23 November 2000. For more information on China's Green Olympic Plan see the official website at www.beijing-2008.org/eolympic/

49 In 1992, China established the China Council for International Cooperation on Environment and Development to function as a government advisory body on the integration of environmental concerns into economic development. The committee (made up of 40 Chinese and international experts) exchanges ideas and provides practical work plans for implementation. However, the influence of the committee, according to the Ford Foundation representative in Beijing, has been piecemeal. Interview with Ford Foundation Representative, Beijing, 10 June 1998. It would seem that the World Bank has had a more visible influence upon Chinese policy-making on the basis of its research programmes in China. This will be discussed in further detail in Chapter 5.

50 In seeking to build a comprehensive legislative framework for the environment, China has imported laws from a number of countries including the United Kingdom, Germany and the United States. In the area of market incentives, China has relied heavily upon the advice of the World Bank.

51 Audrey R. Topping, 'Ecological roulette: damming the Yangtze', *Foreign Affairs*, 74(5), 1995; and Wu Ming, 'Disaster in the making? major problems found in the Three Gorges Dam resettlement', *China Rights Forum: The Journal of Human Rights in China*, Spring, 1998.

52 Wu, 'Disaster in the making?'

53 For critiques on the Three Gorges Dam see Dai Qing, *Yangtze! Yangtze!*, London: Earthscan, 1994; Wu, 'Disaster in the making?'; and Topping, 'Ecological roulette'. For a more official perspective see Shiu-Hung Luk and Joseph Whitney (eds), *Megaproject: A Case Study of China's Three Gorges Dam Project*, Armonk, NY: M.E. Sharpe, 1993.

54 *Renmin Ribao*, 22 July 2002.

55 For a good survey of the project see Liu Changming, 'Environmental issues and the south–north water transfer scheme', *China Quarterly*, 156(December), 1998.

56 Lyman Miller, 'China's leadership transition: the first stage', *China's Leadership Monitor*, 5, 2003.

57 The TransCentury Green Plan is China's blueprint for the environmental goals and priorities referred to in the ninth Five-Year Plan aimed at controlling environmental degradation by the year 2000. The longer term objective behind China's TransCentury Green Plan is to reverse the worsening environmental trends by 2010. China National Environmental Protection Agency, *China's TransCentury Green Plan 1996–2000*, Beijing: Zhongguo huanjing kexue chubanshe, 1997.

58 *China Daily*, 1 November 2002.

59 World Bank, *China, Air, Land, and Water*.

60 Under the reforms the number of government agencies declined from 40 to 29 with a 50 per cent reduction in staff over the three-year period 1998–2001.

61 The director of SEPA, Xie Zhenhua, is reported to have actively negotiated for a 37 per cent decrease in personnel as opposed to the across the board 50 per cent reduction. In the post-restructuring phase, SEPA now has three main departments: international cooperation (global issues and international agreements); foreign economic cooperation (international environmental assistance); and planning and finance (interviews at SEPA, Beijing, 28 April 1999).

62 The National Development and Reform Commission (formerly known as the State Development and Planning Commission) is responsible for macro-level policy-making and long-term planning. It plays a critical role in the integration of economic growth and environmental protection policies. The State Economic and Trade Commission provides an umbrella organization for the former industrial ministries of coal, machinery, metallurgical, chemical, light and textile industries. It is responsible for cleaner production while the Ministry of Science and Technology reinforces the transfer and adoption of environmentally sound technologies. Beijing Centre for Clean Production Newsletter 1998.

63 Numerous studies have assessed the strengths and weaknesses of particular policies. For water pollution discharges, see Abigail Jahiel, 'Policy implementation through organizational learning: the case of water pollution management in China's reforming socialist system', PhD dissertation, University of Michigan, 1994; and for EIAs see Robert Wenger, Wang Huadong and Ma Xiaoying, 'Environmental impact assessment in the People's Republic of China', *Environmental Management*, 14(4), 1990. For *san tongshi* and discharge permits see Barbara Sinkule and Leonard Ortolano, *Implementing Environmental Policy in China*, Westport, CT: Praeger, 1995; and for a recent assessment of the national system of wastewater discharge standards, the pollution discharge fee programme, and the discharge permit system see Xiaoying Ma and Leonard Ortolano, *Environmental Regulation in China: Institutions, Enforcement, and Compliance*, Lanham, MD: Rowman and Littlefield, 2000.

64 Eduard Vermeer, 'Industrial pollution in China and remedial polices', *China Quarterly*, 156(December), 1998.

65 *Beijing Qingnian Bao* (Beijing Youth Daily), 7 November 2000.

66 TVEs can be traced back to the 1950s, when labour-intensive village factories were established in keeping with Mao's policy of rural industrialization or '*litu bu lixiang*' (leave the land but not the village). TVEs proliferated following the dismantling of the commune system and the subsequent shift to a household responsibility system in 1979. See Chan, *Cities with Invisible Walls*; and Guldin (ed.), *Farewell to Peasant China*.

67 By 1997, TVEs accounted for 19 per cent of China's total labour force (28 per cent of the rural labour force) and 29 per cent of total GDP – representing almost one-third of China's economic growth. Xiaolu Wang, 'Economic growth over the past twenty years', in Ross Garnaut and Ligang Song (eds), *China: Twenty Years of Economic Reform*, Canberra: Asia Pacific Press, 1999.

68 However, not all TVEs were simply forced to close down. Instead, four different categories applied – *guan* (complete shutdown), *ting* (stop production and reconsider), *hebing* (merger) and *zhuangqian* (product/service diversification). Interview, Yunnan EPB, 8 June 1999.

69 *China Daily*, 12 August 1997.

70 Vermeer, 'Industrial pollution in China'.

71 Interview, Policy Research Center for Environment and Economy, SEPA, 30 May 1998.

72 Cited in Institute for Human Ecology, *China Environment and Development Report*, 1(1), 1997.

73 It is reported that in August 1999, 22 government and party officials in Shanxi province were removed from office for breaching environmental regulations. FBIS Beijing, *Xinhua*, 5 August 1999. Yet, to date there have been no cases of actual imprisonment.

74 *China Daily*, 21 November 2000.

75 *New York Times*, 16 May 2000.

76 *Zhongguo Qingnian Bao* (China Youth Daily), 4 May 2001.

77 *Zhongguo huanjing bao* (China Environment News), 11 August 1997.

78 Interview, *Renmin Daxue* (People's University), 20 April 1999.

79 Interviews in Beijing April 1999. For a review of China's environmental tax experiment, see Robert A. Bohm, Chazhong Ge, Milton Russell, Jinnan Wang and Jintian Yang, 'Environmental taxes: China's bold initiative', *Environmental Politics*, 40(7), 1998.

80 *China Daily*, 1 June 2002.

81 The world market for environmental equipment and services now exceeds US$54 billion. In China, despite the fact that it is set to become the world's largest market for environmental protection, this sector is still quite small – by 1997 the output value accounted for only 0.7 per cent of China's GDP. Most of the approximately 9,000 enterprises involved were small with low scientific and technological capabilities. *Beijing Review*, 25–31 January 1999. More recently the output value of Chinese environmental industry has increased to 1.9 per cent of GDP. *Renmin Ribao*, 6 November 2001.

82 Established by the International Organization for Standardization in 1996, the purpose behind ISO 14000 is to improve the environmental management systems of industrial enterprises on a worldwide basis.

83 *Renmin Ribao*, 26 November 2001.

84 The role of corporations in environmental protection in China is still at an early stage. For an introductory survey see Jennifer Turner, 'Cultivating environmental NGO–business partnerships', *China Business Review*, November–December, 2003.

85 See OECD, *Environmental Priorities for China's Sustainable Development*, Paris: OECD, 10 December 2001.

86 Civil society can be translated in a number of ways into Chinese including *shimin shehui* (city people's society), *minjian shehui* (people based society), or *wenming shehui* (civilized society). However, *gongmin shehui* (citizen's society) is now in common usage.

87 See Baogang He, *The Dual Role of Semi-Civil Society in Chinese Democracy*, Brighton: Institute of Development Studies, University of Sussex, 1993; Philip C.C. Huang, ' "Public sphere"/"civil society" in China? The third realm between state and society', *Modern China*, 19(2), 1993; and Gordon White, Jude Howell and Shang Xiaoyuan, *In Search of Civil Society: Market Reform and Social Change in Contemporary China*, Oxford: Clarendon Press, 1996.

88 See Anita Chan, 'Revolution or corporatism? Private entrepreneurs as citizens: from Leninism to corporatism', *China Information*, 10(3/4), 1995–1996; Jonathan Unger, ' "Bridges": private business, the Chinese government and the rise of new associations', *China Quarterly*, 147(September), 1996; and Yijiang Ding, 'Corporatism and civil society in China: an overview of the debate in recent years', *China Information*, 12(4), 1998.

89 Tony Saich, 'Negotiating the state: the development of social organizations in China', *China Quarterly*, 161(March), 2000.

90 Official complaints in the form of a letter or visit increased from 142,000 in 1988 to 163,000 in 1996. *Zhongguo tongji nianjain* (China Statistical Yearbook), Beijing: Zhongguo tongji chubanshe, various years.

91 Public dissent over environmental pollution reached a peak in 1994 over the blackening of the Huai River: thousands of paper mills had dumped untreated waste into the river to the extent that the water was no longer even fit for irrigation purposes. In 1996, the central authorities ordered 999 of the paper mills to be closed down. *Newsweek*, 8 October 1996.

92 FBIS Beijing *Xinhua*, Hong Kong Service in Chinese, 5 June 2001. In 1999, China initiated the development of an 'electronic government' as a means of promoting public participation in political decision-making. By the end of 2000, Internet users in China exceeded 20 million.

93 *Beijing Review*, 12 July 1999.

94 Interview, Policy Research Center for Environment and Economy, SEPA, 29 April 1999.

95 *Renmin Ribao*, 4 November 1998; and Human Rights China, *The Emerging Non-Profit Sector: Seeking Independence in a Regulatory Cage*, Beijing: Human Rights China, 13 November 1998.

96 www.fon.org.cn/index.php?id=2684 in Chinese (accessed 10 March 2003).

97 For a comprehensive survey of environmental GONGOs in China see Fenshi Wu, 'New partners or old brothers? GONGOs in transnational environmental advocacy in China', *China Environment Series*, 5, 2002.

98 For a recent study on the role of international support for environmental NGOs working at the grassroots in China see Katherine Morton, 'Transnational advocacy at the grassroots in China: potential benefits and risks', *China Information*, 20(1 and 2), 2006.

99 In her book *The River Runs Black*, Elizabeth Economy has an interesting chapter called 'The new politics of the environment' in which she looks more broadly at the intellectual and political roots of China's environmental movement.

100 Many scholarly works have emphasized the pioneering role of environmental NGOs see Peter Ho, 'Greening without conflict? Environmentalism, NGOs and civil society in China', *Development and Change*, 32(5), 2001.

101 State Council, White Paper, *The Development-Oriented Poverty Reduction Program*, Beijing: State Council, 15 October 2001.

102 Gordon White, *Riding the Tiger: The Politics of Economic Reform in Post-Mao China*, Basingstoke: Macmillan, 1993.

103 See Susan Shirk, *The Logic of Economic Reform in China*, Berkeley, CA: University of California Press, 1993; Shaun Breslin, *China in the 1980s: Centre–Province*

Relations in a Reforming Socialist State, London: Macmillan, 1996; David Goodman (ed.), *China's Provinces in Reform: Class, Community, and Political Culture*, London: Routledge, 1997; and Peter Cheung and Jae Ho Chung (eds), *Provincial Strategies of Economic Reform in Post-Mao China: Leadership, Politics and Implementation*, Armonk, NY: M.E. Sharpe, 1998.

104 Jean C. Oi, 'Fiscal reform and the economic foundations of local state corporatism in China', *World Politics*, 45(1), 1992, pp. 100–1.

105 See Wing-Hung Carlos Lo and Shui-Yan Tang, 'Institutional contexts of environmental management: water pollution control in Guangzhou, China', *Public Administration and Development*, 14, 1994; Hon Chan, Koon-Kwai Wong, K.C. Cheung and Jack Man-Keung Lo, 'The implementation gap in environmental management in China: the case of Guangzhou, Zhengzhou, and Nanjing', *Public Administration Review*, 55(4), 1995; Shui-Yan Tang, Carlos Wing-Hung Lo, Kai-Chee Lo Cheung and Jack Man-Keung, 'Institutional constraints on environmental management in urban China: environmental impact assessment in Guangzhou and Shanghai', *China Quarterly*, 152(December), 1997; and Jahiel, 'The contradictory impact'.

106 Tang *et al.*, 'Institutional constraints'.

107 Interview, Centre for Environmental Sciences, Beijing University, 10 June 1998. In Beihai (designated as China's most unpolluted area with a strong potential for tourism development), the local government recently ignored the EIA and gave approval for a power plant to be constructed right in the centre of the city.

108 Sinkule and Ortolano, *Implementing Environmental Policy*; Tang *et al.*, 'Institutional constraints'; and Abigail Jahiel, 'The organization of environmental protection in China', *China Quarterly*, 156(December), 1998.

109 White, *Riding the Tiger*.

110 A good example is China's administrative reforms initiated in 1998. During visits to Nanning and Wuhan, the younger, more professional and able environmental officials informed me that they were expecting to lose their jobs first, simply for the reason that they had a better chance of finding alternative employment – apparently the vestiges of socialism remain. Interviews, Nanning, 17 June 1999, and Wuhan, 1 July 1999.

111 Hon Chan, K.C. Cheung and Jack M.K. Lo, 'Environmental control in the PRC', in Stuart Nagel and Miriam Mills (eds), *Public Policy in China*, Westport, CT: Greenwood Press, 1993.

112 Jahiel, 'The organization of environmental protection'.

113 *Zhongguo huanjing nianjian 1997* (China Environmental Yearbook 1997), Beijing: Zhongguo huanjing chubanshe, 1997, p. 6.

114 For example, in 1997 the mayor of Shanghai reportedly pledged to invest 3 per cent of the city's GDP in environmental protection. *Beijing Review*, 10–16 March 1997.

115 Interview, SEPA, Beijing, 28 April 1999.

116 *China News Analysis*, 1 December 1994, cited in Elizabeth C. Economy, 'The case of China', Environmental Scarcities, State Capacity, and Civil Violence Project, CISS Occasional Paper, Cambridge, MA: Committee on International Security Studies, American Academy of Arts and Sciences, 1997, p. 11.

117 Hanchen Wang and Liu Bingjiang, 'Policymaking for environmental protection in China', in Michael B. McElroy, Chris P. Nielsen and Peter Lydon (eds), *Energizing China: Reconciling Environmental Protection and Economic Growth*, Cambridge, MA: Harvard University Press, 1998.

118 In 1997, pollution discharge fees collected from TVEs represented only 10 per cent of the national total. Zhang Weijong, Ilan Vertinsky, Terry Ursacki and Peter Nemetz, 'Can China be a clean tiger?: Growth strategies and environmental realities', *Pacific Affairs*, 72(1), 1999.

119 Abigail Jahiel's detailed studies on the implementation of water pollution discharge fees revealed that in Jiangsu province the Xuzhou EPB negotiated an agreement with the local branch of the People's Bank of China for outstanding discharge fees

to be automatically transferred from enterprise accounts into the fee account held by the EPB thereby solving the problem of outstanding payments. In addition, to ensure that returned fees to enterprises were actually being used for environmental protection, Jiangsu province initiated a loan system. See Jahiel, 'The contradictory impact'; and Jahiel, 'Policy implementation'. Both the initiatives were later adopted nationwide.

120 See Chan *et al.*, 'Environmental control in the PRC'; Sinkule and Ortolano, *Implementing Environmental Policy*; Tang *et al.*, 'Institutional constraints'; and Jahiel, 'The contradictory impact'.

121 *Zhongguo huanjing nianjian 1997.*

122 Zhang *et al.*, 'Can China be a clean tiger?'; and Lo and Tang, 'Institutional contexts'.

123 These issues will be addressed in more detail in Chapters 3–5.

124 See State Council, White Paper, 'The development-oriented poverty reduction program'.

125 *China Daily*, 28 July 1999.

3 Engineering a solution: the Japanese approach

1 Speech presented at the International Conference on Financing for Development in Monterrey, 22 March 2002, www.mofa.go.jp/policy/environment/wssd/2002/kinitiative.html (accessed 1 September 2002).

2 By 2001 Japan had funded the establishment of six environmental research and training centres in Thailand, Indonesia, China, Chile, Mexico and Egypt, together with an East Asia acid deposition monitoring network that started operations in 2000. As part of the 'Kyoto Initiative' (presented at the Third Conference of Parties to the United Nations Framework Convention on Climate Change held in Kyoto in 1997), Japan made a commitment to train over 3,000 people from developing countries in air pollution, waste disposal and energy saving technology. More recently, at the World Conference on Sustainable Development in August 2002, Japan pledged to train a further 5,000 people in environment-related activities.

3 See Hiroshi Kanda, 'A big lie: Japan's ODA and environmental policy', *AMPO: Japan–Asia Quarterly Review*, 23(3), 1992; Gavan McCormack, *The Emptiness of Japanese Affluence*, Armonk, NY: M.E. Sharpe, 1996; and Jonathan Taylor, 'Japan's global environmentalism: rhetoric and reality', *Political Geography*, 18(5), 1999.

4 Matsuura Shigenori, 'China's air pollution and Japan's response to it', *International Environmental Affairs*, 7(3), 1995, p. 242.

5 See Peter Evans, 'Japan's green aid', *China Business Review*, July–August, 1994; and Shaun Breslin, *China in the 1980s: Centre–Province Relations in a Reforming Socialist State*, London: Macmillan, 1996.

6 Kusano Atsushi, 'Japan's ODA in the 21st century', *Asia Pacific Review*, 7(1), 2000, p. 47.

7 Japan, Ministry of Foreign Affairs, *Japan's Official Development Assistance Annual Report 1998*, Tokyo: Association for the Promotion of International Cooperation, 1999, p. 33.

8 *Japan Times*, 7 July 2002.

9 At the end of 2000, the anti-China faction within the Liberal Democratic Party (LDP) called for a 30 per cent reduction in aid disbursements to China. Seen as diplomatic suicide by a majority of LDP members this was quickly ruled out by the then foreign minister Kôno Yôhei. After nearly two years of political wrangling in Tokyo, aid to China was eventually cut by 25 per cent in fiscal 2001.

10 The 'military use' of ODA is banned under the 1992 guidelines. But China has allegedly been using sections of expressways built with ODA loans as emergency runways for military aircraft.

11 Traditionally, Japanese aid decision-making has been influenced by four main ministries (*yonshocho taisei*) that included the Economic Planning Agency (EPA). After efforts to streamline the Japanese bureaucracy in 2001, the EPA was abolished and the Ministry of Foreign Affairs assumed the key supervisory role for both grants and loans.

12 This was previously called the Ministry of International Trade and Industry (MITI).

13 On 1 October 2003, JICA changed its status from a public to an independent institution.

14 The merger and its potential implications for the environment will be discussed briefly in Chapter 6.

15 See respectively Dennis Yasutomo, *The Manner of Giving: Strategic Aid and Japanese Foreign Policy*, Lexington, MA: Lexington Books, 1986; Alan Rix, *Japan's Foreign Aid Challenge: Policy Reform and Aid Leadership*, London and New York: Routledge, 1993; and David Arase, *Buying Power: The Political Economy of Japan's Foreign Aid*, Boulder, CO: Lynne Rienner, 1995.

16 Economic cooperation originated in Japan in the 1950s. The principle was established by cabinet order in December 1953 as a means of promoting the private sector with government support. Economic cooperation has also been an important channel for providing war reparations for damages inflicted upon Japan's neighbouring countries during the Second World War. David Arase, 'Public–private sector interest coordination in Japan's ODA', *Pacific Affairs*, 67(2), 1994, p. 173.

17 See Steven W. Hook and Guang Zhang, 'Japan's aid policy since the Cold War: rhetoric and reality', *Asian Survey*, 38(11), 1998.

18 It is not mere coincidence that Indonesia has long been Japan's top recipient of aid because it is Japan's largest supplier of oil outside of the Middle East.

19 METI launched its Green Aid Plan in 1992 as a means of promoting Japanese private sector experience and expertise in environmental protection in developing countries. The budget for 1992–1996 was ¥59 billion (US$500 million).

20 See Miranda Schreurs and Y. Peng, 'The Earth Summit and Japan's initiative in environmental diplomacy', *Futures*, 25(4), 1993; Rowland T. Maddock, 'Japan and global environmental leadership', *Journal of Northeast Asian Studies*, 13(4), 1994; and Susan Pharr and Ming Wan, 'Yen for the earth: Japan's pro-active China environmental policy', in Michael B. McElroy, Chris P. Nielsen and Peter Lydon (eds), *Energizing China: Reconciling Environmental Protection and Economic Growth*, Cambridge, MA: Harvard University Press, 1998.

21 www.mofa.go.jp/policy/environment/wssd/2002/kinitiative.html   (accessed   1 September 2002).

22 Japan, Ministry of Foreign Affairs, *Japan's Official Development Assistance Annual Report 1999*, Tokyo: Association for the Promotion of International Cooperation, 2000.

23 Cited in Japan, Ministry of Foreign Affairs, *Japan's Official Development Assistance Annual Report 1996*, Tokyo: Association for the Promotion of International Cooperation, 1997.

24 Japanese grant assistance is subdivided into technical cooperation (transfer of technology and skills) and grant aid (construction of facilities and procurement of materials and equipment).

25 Cited in Japan, Ministry of Foreign Affairs, *Japan's Official Development Assistance Annual Report 1996*.

26 For detailed descriptions of Japan's pollution cases see Michio Hashimoto, *Economic Development and the Environment: The Japanese Experience*, Tokyo: Ministry of Foreign Affairs, 1992; OECD, DAC, *1994 Environmental Policy Review*, DAC Report, Paris: OECD, 1995; World Bank, *Japan's Experience in Urban Environmental Management*, Washington, DC: Metropolitan Environment Improvement Program, 1995; and Jun Ui, 'Minamata disease and Japan's development', *AMPO: Japan–Asia Quarterly Review*, 27(3), 1997.

27 Ui, 'Minamata disease'.

28 See Peter Dauvergne, *Shadows in the Forest: Japan and the Politics of Timber in Southeast Asia*, Cambridge, MA: MIT Press, 1997. Note here that this relates to the ecological footprint argument advanced in Jim MacNeill, Pieter Winsemius and Taizo Yakushiji, *Beyond Interdependence: The Meshing of the World's Economy and the Earth's Ecology*, New York: Oxford University Press, 1991.

29 Extract from Yoda Susumu, 'Japan's international contribution', keynote speech presented at the International Conference on Energy and Sustainable Development, Tsinghua University, Beijing, 16–17 July 1996.

30 Citizens' environmental protest movements have a long history in Japan dating back to the 1890s when residents of Yantai village protested against toxic wastes from the Ashio Copper mine. The mine was not closed down until 1973. Environmental protests continued intermittently throughout the first half of the nineteenth century and then intensified at the onset of rapid economic growth in the 1960s. Michio Hashimoto, 'Development of environmental policy and its institutional mechanisms of administration and finance', paper presented at International Workshop on Environmental Management for Local and Regional Development, sponsored by UNCRD and UNEP, Nagoya, Japan, 1985.

31 Interview with Dr Michio Hashimoto, Tokyo, 27 March 1999. Dr Hashimoto was the Director of the Pollution Division of the Ministry of Health and Welfare in the 1960s. His untiring efforts to promote pollution control in Japan, despite being branded a communist at the time, has earned him the reputation as Japan's leading authority on the Japanese pollution experience.

32 Katô Kazuo, *The Use of Market-Based Instruments in Japanese Environmental Policy*, Tokyo: Japan Environment Agency, 1993.

33 Japan's four famous cases of pollution-related diseases include the Niigata and Kumamoto Minamata disease as described earlier, Itai-Itai disease resulting from cadmium poisoning from mining companies, and the Yokkaichi asthma caused by the inhalation of sulphur oxides from petroleum refineries.

34 Interview, OECF, Tokyo, 24 March 1999. Hishida Katsuo worked for the Metropolitan Government in Tokyo during the 1960s. He first went to China in 1981 and has subsequently returned 48 times! He has been directly involved in building the human resources capacity of the environmental protection bureau in Chongqing.

35 The literature on environmental assistance to China is surprisingly small. The few studies include Evans, 'Japan's green aid'; Matsuura, 'China's air pollution'; and Pharr and Wan, 'Yen for the earth'.

36 According to official Chinese sources, by 1996 Japanese loans had been responsible for constructing 38 per cent of China's electrified railroads. FBIS Beijing, *Zhongguo Xinwen She* (in Chinese), 25 November 1998.

37 Masato Saito, *Country Study Group for Development Assistance to the People's Republic of China: Basic Strategy for Development Assistance*, Tokyo: JICA, December 1991.

38 The focus areas of cooperation include environmental monitoring, pollution control technology, environmental information, environmental strategy/policy research, and environmental education/public awareness.

39 Before 1994, yen loan packages were negotiated in line with China's five-year plans making it difficult for new aid initiatives to be implemented in the short term. After 1994, loan packages were divided into 3- and 2-year packages. In fiscal 2001, Japan shifted to an annual project-by-project method of disbursement.

40 The fact that at the time the State Development Bank of China had provided US$3.6 billion in funds for the Three Gorges Dam project helped to vindicate the Japanese position, especially as Japanese firms had failed to win any of the procurement contracts. Interview, OECF, Tokyo, 4 September 1997.

41 The Japan Fund for the Global Environment, established in 1993, provides grants to support NGO activities in China. As part of the grant package for fiscal 1997, 18 out

of 30 projects were allocated to China. Fax correspondence, Department of the Japan Fund for Global Environment, Japan Environment Corporation, 14 October 1997. Such activity, however, remains marginal, if not in some cases totally ignored. For example, the Japan Wildlife Association in collaboration with Chinese experts carried out the environmental impact assessment for a JBIC agricultural project in Heilongjiang province. The report was negative because of the perceived potential damaging effect upon wetlands conservation – many birds from Japan migrate to the area. But the project is still going ahead reportedly because the Chinese government's concerns for food security far outweigh its concerns for environmental conservation. Confidential source, 8 April 1998.

42 The expert panel in China consisted of nine members headed by Wang Yangzu, Deputy Director of the State Environmental Protection Administration. In Japan, 14 experts were selected for the panel headed by Professor Toshio Watanabe of the Tokyo Institute of Technology. FBIS Beijing, *Xinhua*, Hong Kong Service (in Chinese) 13 November 1997.

43 Cited in *Nihon Keizei Shimbun*, 3 September 1997 and 4 September 1997; *Japan Times*, 4 September 1997; and *Mainichi Shimbun*, 5 September 1997.

44 In general, JBIC loans are conditioned on a 25–30 year repayment period with a 10 year grace period. Interest rates are varied – 4 per cent for upper middle-income countries, 2.3 per cent for low-middle income countries (including China), and 1 per cent for the poorest countries. Japanese loans remain attractive to developing countries because of their low interest rates, which is a response to concerns over foreign exchange rate fluctuations. In contrast, commercial lending provided by the World Bank's International Bank for Reconstruction and Development has a floating interest rate of around 7.8 per cent.

45 Cited in Motomichi Ikawa, 'JBIC: a new force in global development', *Look Japan*, 'Initiatives for sustainable development: Japan's environmental ODA programs', 45(5), 24 November 1999.

46 Based on the exchange conversion rate of ¥115 to the US dollar in 2001.

47 See JBIC, 'Supporting environmental conservation and human resource development in China – FY2003 ODA loan package for China', www.jbic.go.jp/autocontents/english/news/2004/000025/index.htm (accessed 13 January 2005).

48 The three-model city initiative will be explained in more detail towards the end of this chapter.

49 JICA, 'Report of the second country study for Japan's official development assistance to the People's Republic of China – findings and recommendations', unpublished, Tokyo: JICA, 1999.

50 Japan, Ministry of Foreign Affairs, 'Economic cooperation program for China', www.mofa.go.jp/policy/oda/region/e_asia/china-2.html (accessed 3 March 2002).

51 The tying of aid per se is not a reliable indicator of commercial interests. David Arase makes the point that the United States has a policy of tying aid in its loans to Israel, Egypt and Honduras but these countries are by no means significant economic partners. Arase, 'Public–private sector interest', pp. 171–2. Nevertheless, in those cases when the policy changes from untied to tied loans, as stipulated previously, commercial interests are more clearly a causal factor. At the outset, Japanese loans to China were mainly untied. According to Greg Story, initially Japanese corporations were very successful in winning procurement contracts but later the situation changed, as China became more experienced in administering development assistance and soliciting the interest of a large number of international firms. Greg Story, 'Japan's official development assistance to China: a survey', Pacific Economic Papers No. 150, Canberra: Australia Japan Research Centre, Australian National University, 1987, p. 13. By lowering the interest rates on environmental loans to 0.75 per cent Japan has been able to attach partially tied conditions and still satisfy the OECD guidelines.

52 Partially tied aid is also referred to as less developed country (LDC) untied aid which effectively means that only Japanese firms or developing countries can bid for Japanese loan procurement contracts.

53 According to the Planning and Finance Division of China's SEPA, only 10–20 per cent of OECF procurement contracts go to Japanese companies. Interview, Beijing, 17 June 1998.

54 Interview, Economic Cooperation Bureau, MoFA, Tokyo, 24 March 1999.

55 *Nikkei Weekly*, 8 September 1997.

56 According to Gavan McCormack, more than 1,000 tons of poisonous chemicals and an estimated two million chemical warheads have been found in northeast China since the end of the Second World War. Japan only began a clean up campaign after the international community threatened to nullify Japan's ratification of the United Nations Convention on the Prohibition of Chemical Weapons in 1993. McCormack, *The Emptiness*, p. 264.

57 Interview, Japanese Embassy, Beijing, 7 May 1999.

58 *China Reuters*, 10 August 1997.

59 Matsuura, 'China's air pollution'.

60 Robert Delfs, 'Poison in the sky', *Far Eastern Economic Review*, 156(5), 4 February 1993.

61 *Zhongguo huanjing kexue* (China Environmental Science), October 1997.

62 *Japan Times*, 26 January 2002.

63 Ryukichi Imai, 'Global governance: some reflections', IIPS Policy Paper 191E, Tokyo: Institute for International Policy Studies, December 1997, p. 27.

64 Note here that negotiations were also stalled following China's nuclear testing in 1995.

65 Interviews, OECF, Tokyo, 4 September 1997, and OECF, Beijing, 27 May 1998.

66 See James Feinerman, 'Chinese participation in the international legal order: rogue elephant or team player?', *China Quarterly*, 141(March), 1995; and Shaun Breslin, 'China's environmental crisis in a global context', *Global Society: Journal of Interdisciplinary International Relations*, 10(2), 1996.

67 Japan–China Expert Committee, 'Nitchû kankyô kaihatsu moderu toshi kôsô ni kansuru teigen' (Policy recommendations for the Japan–China environmental model city project), unpublished draft report, Tokyo: JICA, 1998.

68 ODA loans for fiscal 2000 included the Beijing environmental improvement project that aims to set up Beijing's first natural gas cogeneration facilities and the Ningxia afforestation and vegetation cover project that is designed to prevent desertification by increasing the amount of vegetation cover. See www.jbic.go.jp (accessed 28 November 2002).

69 One might even argue that a coal transportation project is both more economically and environmentally viable because of the high sulphur content of coal deposits in the South.

70 It should be noted here that under JBIC guidelines, loans provided for capital construction projects are not to be used to cover maintenance or operating costs.

71 Interview, Planning and Finance Division, SEPA, Beijing, 17 June 1998.

72 Interview, Chief Representative, OECF (JBIC), 27 May 1998.

73 From 1992 to 1998, the Japanese private sector under METI's Green Aid Plan funded 25 environmental projects in China at a total cost of US$182 million. Out of the total number of projects, 17 focused on energy conservation and clean coal technology, and 4 on simplified flue gas desulphurization (whereby the technology used has a low desulphurization rate of 70 per cent but relatively low installation costs). Interview, New Energy and Industrial Technology Development Organization (NEDO), Beijing, 19 June 1998.

74 Ibid.

75 Interview, National Development and Reform Commission, Beijing, 4 June 1998.

76 Interview, Economic Cooperation Division, MITI, Tokyo, 25 March 1999.

77 Interview, JICA, Tokyo, 26 March 1999.

78 Interview, OECF, Tokyo, 4 September 1997.

79 Interview, Planning Department, JICA, Tokyo, 26 March 1999.
80 Interview, OECF, Tokyo, 23 March 1999.
81 Interviews, deputy-director of Guangxi Regional EPB, 7 July 1998; director, Huhhot EPB, 16 July 1998; and deputy-director, Jilin Provincial EPB, in Beijing, 24 June 1998.
82 Note here that by 1999 JBIC had a total of nine environmental improvement projects. The four projects selected here were amongst the earliest.
83 OECF, Quarterly Report, *Outline of 40 Planned Projects for Fourth Batch of ODA Loans to China*, Tokyo: OECF, August 1997.
84 Based upon a comparison of the Interim Report for OECF, Special Assistance for Project Formation Study, 'People's Republic of China environmental improvement project – interim report of China environmental improvement project', unpublished, Tokyo: OECF, March 1995; and EPB reports (unpublished) in Huhhot, Liuzhou and Benxi.
85 The local coordination of environmental problems in general involves at least five local bureaucracies. The EPB is responsible for environmental regulations, taxes and hazardous waste, the Planning Committee for energy conservation, the Urban Construction Commission for sewage and solid waste, the Bureau of Public Affairs for water supply, and the Economic Trade Committee for cleaner production.
86 It is estimated that 68 per cent of industry in the province belongs to the state and 5.6 million out of 9.1 million workers are employed by SOEs. Over 50 per cent of Shenyang's working population is employed in heavy industry such as iron and steel, coal mining, smelting and petrochemicals. Louise Cadieux, 'Liaoning in transition of reform: what will become of the state-owned enterprises', *China Today*, 48(11), 1999.
87 Ibid.
88 *Shenyang chengshi ke chixu fazhan xiangmu: ke chixu fazhan de jintian he mingtian* (Shenyang Sustainable City Development Project: sustainable development today and tomorrow), Shenyang: Huanjing baohu ju, 1997.
89 Shenyang Planning Committee, 'Shenyang shi liyong riyuan daikuan shishi daqi wuran zhili xiangmu qingkuang jieshao' (Introduction to Shenyang City implementation of the Japanese yen loan project to control air pollution), unpublished, Shenyang: Shenyang Green Project Office, 1999.
90 Shenyang has conducted international exchanges with more than 20 countries and inter-national organizations. Shenyang Environmental Protection Bureau, 'The profile of Shenyang environmental protection', unpublished, Shenyang: Shenyang EPB, May 1999.
91 At the initiation of the project, institutional arrangements were not firmly in place. Shenyang was the first city to receive JBIC environmental assistance and the Planning Committee seized control in the belief that the project was more about energy conservation than pollution control. At the request of the Japanese government, subsequent projects were placed under the control of the municipal EPBs. Interview, Shenyang Green Project Office, 12 May 1999.
92 Shenyang Planning Committee, 'Shenyang shi liyong riyuan daikuan'.
93 Interviews, Guiyang EPB, 14 June 1999, and Liuzhou EPB, 22 June 1999.
94 Interview, Shenyang Project Office, 12 May 1999.
95 Ibid.
96 Ibid.
97 Ibid.
98 *Huasheng Monthly*, 9 January 2001.
99 See JBIC, 'List of anti-global warming projects (FY1998–FY2003)', www.jbic.go.jp/english/environ/support/overseas/warming.php (accessed 21 March 2005).
100 OECF, Quarterly Report, *Outline of 40 Planned Projects*.
101 Interview, Benxi Agenda 21 Office, 17 May 1999.
102 Interview, Japanese loan project office, Benxi EPB, 18 May 1999.
103 Ibid.
104 Ibid.

105 Ibid.

106 Before 1949, the region's two industrial cities – Huhhot and Baotou – were largely semi-desert and inhabited by Mongol herders. After 1949, roads and railways were constructed to facilitate access to the region's mineral deposits, which stimulated the beginnings of industrialization. Industry now accounts for 40 per cent of the region's economic output. *Zhongguo tongji nianjian 1997* (China Statistical Yearbook 1997), Beijing: Zhongguo tongji chubanshe, 1997, p. 51.

107 Interview, Director, Huhhot EPB, 16 July 1998.

108 The OECF also provided a US$94 million loan to the city of Baotou.

109 Interviews, Director, Huhhot EPB, 16 July 1998, and Director, Foreign Economic Cooperation Office, Inner Mongolia Regional EPB, 17 July 1998.

110 Interview, deputy-director, Guangxi Regional Environmental Protection Bureau, 7 July 1998.

111 Liuzhou Environmental Protection Bureau, 'The situation of ambient air pollution in Liuzhou', unpublished, Liuzhou: EPB, 1998. Note here that during the mid-1990s regulations had been introduced to enforce the use of higher grade coal in industrial enterprises. For example, the Liuzhou Chemical Fertilizer Plant switched to using Shaanxi coal which is smokeless and has a sulphur content of only 0.5 per cent. Interview, Liuzhou Chemical Fertilizer Plant, 9 July 1998.

112 This situation is likely to change in the future. In 1999 the Liuzhou Power Plant was using technologies sourced from Japanese companies under partially tied loan conditions.

113 Interview, Liuzhou Iron and Steel Plant, 23 June 1999.

114 Interview, Liuzhou Chemical Fertilizer Plant, 23 June 1999.

115 Interview, Liuzhou EPB, 22 June 1999.

116 Interview, Director, Liuzhou EPB, 22 June 1999.

117 Cited in *Nikkei Weekly*, 11 January 1999.

118 Interview, Division for International Cooperation, SEPA, 26 June 1998.

119 Dalian Environmental Protection Bureau, 'Promotion of the construction plan of Dalian environmental demonstration zone: towards sustainable development', unpublished, Dalian: Dalian EPB, 1998.

120 Hoshina Hideaki, 'Dairen-shi toshi kankyô moderu chika keikaku chôsa' (Basic research plan on Dalian environmental model city project), unpublished draft report, Tokyo: JICA, 1999.

121 Interview, Director, Dalian Model Project, JICA, 23 March 1999.

122 Interview, Dalian Cement Factory, 25 May 1999.

123 Interview, Dalian Project Office, 24 May 1999.

124 Interview, Dalian Cement Factory, 25 May 1999.

125 Interview, Guiyang EPB, 14 June 1999.

126 *Zhongguo tongji nianjian 1997*. Total investment resources include the local government budget for capital construction, renewal, and transformation, profits from the comprehensive utilization of wastes, environmental protection subsidies, environmental protection loans and foreign capital.

127 Interview, Guiyang Steelworks, 14 June 1999.

128 See Yiping Huang, 'State-owned enterprise reform', in Ross Garnaut and Ligang Song (eds), *China: Twenty Years of Economic Reform*, Canberra: NCDS Asia Pacific Press, 1999; and Barry Naughton, 'China: domestic restructuring and a new role in Asia', in T.J. Pempel (ed.), *The Politics of the Asian Economic Crisis*, Ithaca, NY: Cornell University Press, 1999.

4 Managing the environment with a human face: the UNDP approach

1 Aside from a study by Poul Engberg-Pederson and Claus Hvashøj Jørgensen, 'UNDP and global environmental problems: the need for capacity development at country

level', in Helge Ole Bergesen and Georg Parmann (eds), *Green Globe Yearbook of International Co-operation on Environment and Development*, Oxford: Oxford University Press, 1997, the most recent studies are in the form of UNDP reports related to specific countries. For a general overview, albeit one that is now out of date, see UNDP, *Protecting the Environment*, New York: UNDP, 1988.

2 For a good overview of the UNDP's changing mandate see Alexander Timoshenko and Mark Berman, 'The United Nations Environment Programme and the United Nations Development Programme', in Jacob Werksman (ed.), *Greening International Institutions*, London: Earthscan, 1996.

3 This objective was advanced in John McHale and Magda McHale, *Basic Human Needs*, New Brunswick, NJ: Transaction Books for UNEP, 1978; and Paul Streeten, Shahid Javed Burki, Mahbub ul-Haq, Norman Hicks and Frances Stewart, *First Things First*, Oxford: Oxford University Press, 1981.

4 Mostafa Tolba, *Development without Destruction: Evolving Environmental Perceptions*, Dublin: Tycooly, 1982; and World Commission on Environment and Development, *Our Common Future*, Oxford: Oxford University Press, 1987.

5 Richard Jolly, Giovanni Cornia and Frances Steward (eds), *Adjustment with a Human Face*, Oxford: Clarendon Press, 1989.

6 Agenda 21 was adopted by the international community at the UNCED in 1992 with a set of normative principles to promote an economically, socially and environmentally sustainable world economy. To this end, Agenda 21 calls for the burden of responsibility to be borne by all, and not disproportionately by the poor. For an overview of Agenda 21 see Paul Sitarz and Daniel Sitarz (eds), *Agenda 21: The Earth Summit Strategy to Save Our Planet*, Boulder, CO: Earth Press, 1994. While some scholars have supported the document for its symbolic significance others have been highly critical in dismissing it as a weak compromise with little meaning or influence in practice. See Pertap Chatterjee and Matthias Finger, *The Earth Brokers: Power, Politics, and World Development*, London and New York: Routledge, 1994.

7 Speech by James Gustav Speth, Administrator UNDP, International Conference on Population and Development, Cairo, 6 September 1994.

8 Engberg-Pederson and Jørgensen, 'UNDP and global environmental problems'.

9 In order to improve coordination and policy coherence a United Nations Development Group was established in 1997 comprising the executive heads of all UN development agencies and funds (including the World Food Program, the UN Centre for Human Settlements (UNCHS), UN Children's Fund, and the UN Population Fund amongst others) under the leadership of the UNDP Administrator. See UNDP, *Annual Report 1996/1997, Ending Poverty and Building Peace through Sustainable Human Development*, New York: UNDP, October 1997.

10 UNDP, *Annual Report 2003, A World of Development Experience*, New York: UNDP, 2003.

11 Ibid.

12 UNDP, *Annual Report 1996/1997*.

13 UNDP and UNPF, *The Results-Oriented Annual Report 2001*, New York: UNDP, 2001.

14 The Millennium goals include eradicating extreme poverty, achieving universal primary education, promoting gender equality and empowering women, reducing child mortality, combating HIV/AIDS, ensuring environmental sustainability, and developing a global partnership for development by 2015. For full details see http://ddp-ext.worldbank.org/ext/MDG/home.do

15 UNDP, *Annual Report 2003*, p. 1.

16 UNDP, *UNDP Today: Introducing the Organisation*, New York: UNDP, September 1998; and UNDP and UNPF, *Results-Oriented Annual Report (ROAR)*, New York: UNDP, 2000.

17 UNDP, *Annual Report 1996/1997*; and UNDP, *Annual Report, Partnerships to Fight Poverty*, New York: UNDP, 2001, p. 18.

18 UNDP, *Capacity 21 Annual Report*, New York: UNDP, 2001.

19 Ibid.

20 The Montreal Protocol on Substances that Deplete the Ozone Layer was signed in 1987 with further amendments to the Protocol made in London in 1990, Copenhagen in 1992 and Vienna in 1995. The Protocol sets out a time frame for phasing out and reducing ODS. Developed countries eliminated halon consumption in 1994 and chlorofluorocarbon (CFC) consumption in 1996. Developing countries have a grace period up until 2010.

21 Interview, Energy and Atmosphere Programme, Sustainable Energy and Environment Division, UNDP, New York, 30 March 1999.

22 For further details of the UNDP's focus upon energy and the environment see José Goldenberg and Thomas Johannson, *Energy as an Instrument for Socio-Economic Development*, New York: UNDP, 1995; and UNDP, *UNDP Initiative for Sustainable Energy*, New York: UNDP, June 1996. More recently, the UNDP together with the UN Department of Economic and Social Affairs (UNDESA) and the World Energy Council have launched two reports on the links between energy, poverty and development. See UNDP and UNDESA, *World Energy Assessment: Energy*, New York: UNDP, 2000; and UNDP and UNDESA, *The Challenge of Sustainability*, New York: UNDP, 2000.

23 The UNDP also draws upon the notion of sustainable livelihoods in its environmental assistance. This parallel theme is primarily applied to rural rather than urban settings. The idea of sustainable livelihoods surfaced in the late 1980s and is concerned with securing access to assets, maintaining resource productivity and ensuring adequate cash to meet basic needs. It is of particular relevance to the UNDP's poverty eradication programmes. For an overview of the concept see Robert Chambers and Gordon Conway, 'Sustainable rural livelihoods: practical concepts for the twenty-first century', IDS Discussion Paper 296, Brighton: Institute of Development Studies, 1992; and Goran Hyden, 'Governance and sustainable livelihoods: challenges and opportunities', paper presented at Workshop on Sustainable Livelihoods and Sustainable Development, University of Florida, 1–3 October 1998.

24 See UNDP, *Empowering People: A Guide to Participation*, New York: UNDP, 1997.

25 Interview, Social Development and Poverty Alleviation Division, UNDP, New York, 31 March 1999.

26 'Dangdai Zhongguo' cong shu bian ji bu (ed.), *Dangdai zhongguo de duiwai jingji hezuo* (Contemporary Chinese Economic Cooperation with Foreign Countries), Beijing: Zhongguo shehui kexue chubanshe, 1989.

27 Engberg-Pederson and Jørgensen, 'UNDP and global environmental problems'.

28 Interview, Energy and Atmosphere Programme, Sustainable Energy and Environment Division, UNDP, New York, 30 March 1999.

29 UNDP China website, www.edu.cn/undp/ccp/cp4/ccf/ccfintro.htm (accessed 15 March 1999.

30 Ibid.

31 UNDP China, *China Environment and Sustainable Development Resource Book II: A Compendium of Donor Activities,* Beijing: UNDP, 1996.

32 UNDP, 'GEF portfolio for China', unpublished, New York: UNDP, March 1999; and www.undp.org/seed/eap/montreal/index.htm (accessed 13 March 1999).

33 Calculated from UNDP China, *Environment and Sustainable Energy Development: A Strategy and Action Plan for UNDP in China*, Beijing: UNDP, November 1997; UNDP, 'GEF portfolio for China'; and UNDP, *The Fifth Year: Learning and Growing*, Annual Report Capacity 21, New York: UNDP, 1998.

34 UNDP and UNPF, *Second Country Cooperative Framework for China (2001–2005)*, New York: UNDP, September 2001, p. 8.

35 Interview, UNDP, Beijing, 2 June 1998.

36 MOFTEC's role in managing technical assistance to China sits somewhat uncomfortably with its central role in the promotion of trade and private investment activities. After China's opening to the world in 1979, MOFTEC was given the responsibility for grant related development assistance for the simple reason that MOFTEC officials, unlike their counterparts in the former State Planning Commission, could communicate effectively in English.
37 The SCP will be discussed in more detail in the second half of this chapter.
38 Interview, UNIDO, Beijing, 2 July 1998.
39 Interview, Energy and Atmosphere Programme, Sustainable Energy and Environment Division, UNDP, New York, 30 March 1999.
40 State Council, *China's Agenda 21: White Paper on China's Population, Environment, and Development in the 21st Century*, Beijing: Zhongguo huanjing kexue chubanshe, 1994.
41 Ibid., p. 223.
42 UNDP China, www.edu.cn/undp/news/htm (accessed 25 November 1999), my emphasis.
43 Aside from progress at the national level, efforts under the guidance of the UNDP are also underway to 'localize' China's Agenda 21. In 1999 three provinces had been selected as pilot sites together with 10 cities and 30 experimental sustainable communities such as Xitan district in Beijing. These pilot projects are supported by the Administrative Centre for China's Agenda 21 (ACCA21) in Beijing. Interview, ACCA21 Office, Beijing, 29 April 1999.
44 Interview, Capacity 21 Office, New York, 1 April 1999.
45 Interview, Energy and Atmosphere Programme, Sustainable Energy and Environment Division, New York, 30 March 1999.
46 CICETE, *China/UNDP The Fourth China Country Programme: Environment and Energy Projects*, Beijing: CICETE, 1996.
47 This trend has continued with more recent environmental and energy projects targeting the cities of Liuzhou, Meishan and Taiyuan as well as rural areas.
48 The SCP was launched in 1990 and was later recognized at UNCED as an important vehicle for implementing Agenda 21 at the municipal level. By 2000, 20 cities worldwide were participating in the SCP including the two cities in China.
49 Habitat/UNEP, *Sustainable Cities Programme: Approach and Implementation*, Nairobi: Habitat/UNEP, 1998.
50 For further details of the SCP approach see Habitat/UNEP, *The SCP Process Activities: A Snapshot of What They Are and How They Are Implemented*, Nairobi: Habitat/UNEP, 1998; and Habitat/UNEP, *SCP 1997 Meeting Report*, Nairobi: Habitat/UNEP, 1998.
51 At the national level, the Ministry of Construction has overall authority over human settlements including urban planning, infrastructure, water supply and drainage, solid waste management, sanitation, urban traffic, heating and gas, energy efficiency, alternative energy sources and national disaster alleviation.
52 UNDP China, 'Managing sustainable development in Shenyang', Project No. CPR/96/321/A/01/99, unpublished, Beijing: UNDP, 1996, p. 4.
53 Ibid.
54 ACCA21 was set up under the direction of the Ministry of Science and Technology and the former State Planning and Development Commission to coordinate and implement sustainable development activities at the national and sub-national levels.
55 Other key agencies included the Municipal Economic Restructuring Commission, Civil Affairs Bureau, Public Health Bureau, Finance Bureau, Land Planning Bureau and the municipal branch of the People's Bank of China.
56 Interview, Shenyang Project Office, Shenyang, 10 May 1999.
57 To this end UNEP, in collaboration with the WHO and the World Bank provided technical support by establishing four air monitoring stations integrated with UNEP's Global Environmental Monitoring System.

58 Discussion based upon interviews at the Shenyang Project Office, 10 May 1999.

59 *Shenyang chengshi ke chixu fazhan xiangmu: ke chixu fazhan de jintian he mingtian* (Shenyang Sustainable City Project: sustainable development today and tomorrow), Shenyang: Huanjing baohu ju, 1997.

60 Shenyang Sustainable City Project, *Shenyang Environmental Profile*, Shenyang: Sustainable Shenyang Project Office, April 1998, p. 67.

61 Shenyang Sustainable City Project, *Chengshi zixun dahui. Daibiao shouce* (City consultation conference: participants manual), Shenyang: Sustainable Shenyang Project office, 5–7 May 1998.

62 Interview, Shenyang Project Office, 10 May 1999.

63 Interview, Deputy Project Manager, Shenyang Project Office, 14 May 1999.

64 FBIS *Shenyang Liaoning Ribao*, 3 August 1997.

65 At the annual meeting of the SCP in Shenyang (attended by over 100 participants from around the world) the SCP partners pledged to work together to build a strong SCP at the global level through information sharing and the pooling of resources. The pledge was formalized in the so-called 'Shenyang Declaration'. Habitat/UNEP, *SCP 1997 Meeting Report*.

66 Shenyang Environmental Protection Bureau, 'The profile of Shenyang environmental protection', unpublished, Shenyang: Shenyang EPB, May 1999, p. 2.

67 Interview, Shenyang Project Office, 14 May 1999.

68 A 'complaints hotline' now exists in every local government department together with a general hotline (#2211) which was allegedly the first in China.

69 Shenyang Environmental Protection Bureau, 'The profile of Shenyang environmental protection'.

70 Interview, Shenyang Project Office, 14 May 1999.

71 Ibid.

72 At the time of my visit to Shenyang, the European project was still suffering severe delays in project implementation, which were more reflective of problems in Brussels than in Shenyang. Interview, Environment Representative, European Commission Delegation Office, Beijing, 27 April 1999.

73 Interview, European Union Project Office, Liaoning EPB, Shenyang, 11 May 1999.

74 Interview, Shenyang Project Office, 14 May 1999.

75 Leading Office for Implementing China's Agenda 21, *Wuhan Environment Profile*, Wuhan: Agenda 21 Office, 1998.

76 Zhang Naidi, 'Automobile exhaust pollution and its control in Wuhan', in Leading Office for Implementing China's Agenda 21, *Wuhan Environment Profile*.

77 Leading Office for Implementing China's Agenda 21, *Wuhan Environment Profile*.

78 Flooding in Wuhan has a long history. In 1954, in the worst floods on record, the Yangzte River reached almost 30 metres causing the death of 85,000 people. Leading Office for Implementing China's Agenda 21, *Wuhan Environment Profile*.

79 Ibid.

80 By 1999, Wuhan had received environmental assistance from the World Bank (for the treatment of domestic sewage and clean up of the Donghu Lake), the Netherlands (for a landfill site), Finland (for wastewater treatment) and Japan (for flood control).

81 The solid waste problem in Wuhan is no worse than most other cities in China. There is no obvious reason why it should be selected as an environmental priority except for the fact that in the past the Wuhan government has tended to focus more upon disposal rather than collection. Interview, Habitat advisor, Wuhan, 2 July 1999.

82 Interview, Wuhan Project Office, Wuhan, 1 July 1999.

83 Hubei is one of China's three pilot provinces for implementing China's Agenda 21 together with Hebei and Shanxi. The leading group for implementing China's Agenda

21 in Wuhan is comprised of over 20 different government agencies and its office is located in the Planning Committee rather than the EPB.

84 Interview, Wuhan Project Office, 3 July 1999.

85 Leading Group Office for Implementing China's Agenda 21, *1998 Wuhan chengshi ke chixu fazhan huanjing wenti zixun dahui* (Sustainable Development Environmental Issues Consultation Conference in Wuhan 28–30 April 1998), Wuhan: Wuhan Planning Committee, 1998.

86 Interview, Wuhan Project Office, 4 July 1999.

87 One example was a project to centralize small snack stalls into one area so that they could be controlled more easily. Government at the district level provided the funds with an additional US$2,000 from the EPB.

88 In October 1999, FETC, with support from the SETC in Beijing, held a conference in Hamburg on foreign investment for environmental protection in Wuhan. According to one FETC official, overall private investment in environmental protection was still limited because investors were risk adverse and sufficient policy guidelines were not in place. Despite these constraints, the municipal government was in the process of considering the city's first Build Operate and Transfer investment project for a solid waste incinerator in collaboration with a company in France. Interview, FETC, Wuhan, 2 July 1999.

89 Email correspondence with Wuhan Project Office, 28 February 2000.

90 Report provided in confidence from the UNCHS, Nairobi, 2000.

91 Email correspondence with Habitat advisor to the SCP projects in China, 6 April 2002.

92 Interview, Centre for Environmental Sciences, Peking University, Beijing, 30 April 1999.

93 Benxi's Agenda 21 Leading Group Office, 'Benxi Agenda 21: overall design for the sustainable development strategy of Benxi City', unpublished, Benxi, 1994.

94 Benxi was not the first city in China to focus upon cleaner production, although it remains amongst a small minority. National efforts to promote cleaner production began in earnest in 1993 when the World Bank provided a US$6.2 million loan to SEPA to develop and test an approach to cleaner production in China with technical support from the Industry and Environment Office of UNEP. Pilot audits for the adoption of cleaner production techniques were carried out amongst enterprises in Beijing, Shaoxin and Yantai. In 1994, Benxi initiated a cleaner production plan involving 11 pilot enterprises but, according to the UNDP, the city's capacity to conduct environmental audits based upon costs and technical feasibility was limited. UNDP China, 'Capacity building for widespread adoption of clean production for air pollution control in Benxi', Project No. CPR/96/307/B/01/99, unpublished, Beijing: UNDP, 1996, p. 11.

95 Ibid.

96 The National Center for Cleaner Production was established in 1994 to promote the coordination and dissemination of cleaner production techniques in China. It was funded by the Asian Development Bank and is located in the ACCA21 Office. A duplication of donor assistance appears to have taken place in that the UNDP had previously established a cleaner production centre at CRAES in 1993.

97 Interview, Benxi Agenda 21 Office, 17 May 1999.

98 Interview, Director, Environment Division, Benxi Cement Factory, 18 May 1999.

99 Interview, Director, Environment Division, Beigang Cast Iron Steelworks, 18 May 1999.

100 The environmental audit of the 15 enterprises had been funded through polluter fees but this source of funding was clearly limited.

101 Following decades of unlimited extraction, it is estimated that only a 70-year supply of iron ore remains in Benxi. Coal reserves are also extremely scarce. Benxi is now importing iron ore from Australia and coal resources have to be imported domestically.

102 A government opinion poll carried out in 1989, with the aim of identifying the worst sources of pollution in Benxi, involved 10,000 residents.

103 Interview, Benxi Agenda 21 Office, 17 May 1999.

104 CICETE, *UNDP Air Pollution Control Programme*, Beijing: CICETE, 1996.

105 UNDP China, 'Capacity development for acid rain and SO_2 pollution control in Guiyang', Project No. CPR/96/304/A/01/99, unpublished, Beijing: UNDP, 1996.

106 The problem is that regulatory controls in Guiyang do not provide the necessary incentive for enterprises to reduce SO_2 emissions. In effect, a fee system needs to be established which reflects the costs of pollution control.

107 Interview, Acid Rain Center, Guiyang EPB, Guiyang, 14 June 1999.

108 UNDP China, 'Summary report of the evaluation mission: China urban air pollution control program', Beijing: UNDP, May 2001.

109 The Centre for Environmental Sciences at Peking University is somewhat of an aberration. Unlike the majority of environmental science institutes in China, the Centre carries out scientific and economic research. Economists working at the Centre focus chiefly upon cost-effective approaches to environmental management based upon the theories of environmental economics.

110 Interview, Acid Rain Center, Guiyang EPB, Guiyang, 14 June 1999.

111 Interviews in Beijing at CICETE, 29 April 1999, and CRAES, 30 April 1999.

112 UNDP China, 'Summary report of the evaluation mission'.

113 Freidrich Ernst Schumacher, *Small is Beautiful: A Study of Economics as if People Mattered*, London: Blond and Briggs, 1973, p. 31.

114 UNDP China, 'Summary report of the evaluation mission'.

115 This is likely to change in the future in view of planned administrative restructuring at the local level. There have been rumours that the Agenda 21 Office will be placed under the authority of the EPB. At the time of the implementation of the project, Madam Tang Guimei was opposed to this happening because in her mind the EPB was too narrowly focused upon enforcing regulations rather than working with enterprises to arrive at a solution. Fax correspondence with Madam Tang Guimei, 20 March 1999.

116 UNDP China, 'Summary report of the evaluation mission'. The report also reveals that the EPB director who was originally responsible for project implementation tragically died – ironically of lung cancer.

117 Interview, ACCA21 Office, Beijing, 29 April 1999.

118 By May 1999, the project offices in Shenyang and Wuhan were negotiating with the UNDP and Habitat over follow-up funds for implementing environmental plans. The project office in Benxi was requesting funds to set up a Cleaner Production Centre, which the UNDP were likely to refuse on the basis that if the profits from cleaner production were so high, the Centre should be self-financing. The UNDP was understandably cautious about providing extra funds for Benxi in light of the fact that they were still awaiting cost-sharing funds from the Benxi municipal government. Interviews, UNDP, Beijing, 5 July 1999.

119 Interview, Habitat advisor, Wuhan, 2 July 1999.

120 Ibid.

5 Creating incentives and institutions: the World Bank approach

1 The Bank is also trying to implement a participatory approach to natural resource management in China through its Sustainable Forestry Development Project initiated in 2002.

2 The IFC was established in 1956 to provide financial support to private enterprises, the IDA was created in 1960 to provide concessional funding to low-income countries and the ICSID was set up in 1965 to facilitate foreign investment in developing countries by providing conciliation and arbitration facilities. Some years later in

1988, MIGA was established to support investment flows to developing countries by securing guarantees against non-commercial risks to foreign investors on behalf of its members.

3 Legally the Bank is required to obtain a government guarantee in cases where the borrower is not the government. Ibrahim Shihata, *The World Bank in a Changing World*, Dordrecht: Martinus Nijhoff Publishers, 1991, p. 8.

4 These credits, however, do carry a service charge of less than 1 per cent.

5 The World Bank Group, including IFC, MIGA and ICSID, has 14,816 employees, of whom 10,205 are professional staff. Email correspondence with Bank employee, 5 May 2000.

6 See Bruce M. Rich, *Mortgaging The Earth: The World Bank, Environmental Impoverishment, and the Crisis of Development*, Boston, MA: Beacon Press, 1994; Kevin Danaher (ed.), *50 Years is Enough: The Case Against the World Bank and the International Monetary Fund*, Boston, MA: Southend Press, 1994; and Catherine Caufield, *Masters of Illusion: The World Bank and the Poverty of Nations*, London: Pan Books, 1998. Many of the Bank's most vociferous environmental critics quite often (but certainly not always) are highly sceptical of the purpose of development aid per se. In such cases, strong moral principles are at stake over the bigger question of whether richer countries should be in the business of providing financial transfers to poorer countries, which arguably do little more than increase Third World debt. Or, in the same vein, whether richer countries should be providing environmental loans which encourage recipient countries to exploit their natural resources in unsustainable ways in order to service the debt. Such criticism, however, remains a largely Northern-based agenda. Very few NGOs in developing countries, if any, have questioned the Bank's very existence for fear of jeopardizing capital flows to their countries. Moreover, the dominant cry from poorer countries worldwide has been for more environmental funds from developed nations, not less.

7 The United States carried this out instead on a bilateral basis through the Marshall Plan initiated in 1947.

8 Structural adjustment lending was a response to the Third World debt crisis in the 1980s. The basic objective was to pressure indebted countries to reduce domestic expenditures and export more in order to earn more foreign exchange to service their debts. The social impact was devastating: real incomes declined and health and education services suffered as a consequence. For damaging critiques of the social and environmental impact of structural adjustment lending, see Rich, *Mortgaging the Earth*; and David Reed, *Structural Adjustment and the Environment*, Boulder, CO: Westview Press, 1992. Paul Mosley, Jane Harrigan and John Toye, *Aid and Power: The World Bank and Policy-Based Lending*, volumes 1 and 2, London: Routledge, 1991, provide a detailed assessment of the largely negative influence of structural adjustment upon economic growth and aggregate investment. For a penetrating critique of the impact of structural adjustment upon Third World debt see Susan George and Fabrizio Sabelli, *Faith and Credit: The Bank's Secular Empire*, Washington, DC: Institute for Policy Studies, 1994.

Sectoral adjustment lending is when disbursements are made to achieve overall sector goals, in energy or railways for example, rather than specific projects. It became particularly prevalent in the 1980s but has received far less attention since.

9 Nicholas Stern and Francisco Ferreira, 'The World Bank as "intellectual actor" ', in Devesh Kapur, John Lewis and Richard Webb (eds), *The World Bank: Its First Half Century*, volume 2, Washington, DC: Brookings Institution, 1997.

10 Interview, Social Development Division, World Bank, Washington, DC, 7 April 1999.

11 The former president of the Bank, James Wolfensohn, first presented the new policy on development partnerships to the Bank's Board of Governors in 1998. See James Wolfensohn, 'The other crisis', address to the Board of Governors, Washington, DC: World Bank, 6 October 1998.

12 In 2003 seven key focal areas were identified including environmental sustainability, water supply and sanitation, education for all, HIV/AIDS, maternal and child health, investment climate and finance, and trade.

13 World Bank, *World Bank Annual Report 2003*, Washington, DC: World Bank, 2003.

14 On a wider scale the net negative transfer between developing countries and their public and private creditors led to the Third World debt crisis in the 1980s. In 1982, according to Rich, *Mortgaging the Earth*, p. 109, Latin American countries borrowed US$49.63 billion in total and paid back US$66.81 billion leading to a net negative transfer of US$17 billion.

15 Edward Mason and Robert Asher, *The World Bank since Bretton Woods*, Washington, DC: Brookings Institution, 1973.

16 The poor performance of World Bank projects was stressed in an internally commissioned report led by the then vice-president Willi Wapenhams in 1992. The report highlighted the Bank's systemic failure to ensure project outcomes – with 37.5 per cent of total projects at the time rated as a failure. This was based largely upon rates of economic return. Obviously the failure rate would have been considerably higher if assessment had been based upon environmental and social returns on investment. See Willi Wapenhams, *Report of the Portfolio Management Task Force*, Washington, DC: World Bank, 1992.

17 Shihata, *The World Bank in a Changing World*.

18 Ibid.

19 Mick Moore, 'Toward a useful consensus', *IDS Bulletin*, 29(2), 1998, p. 45.

20 World Bank, *World Bank Annual Report 2003*.

21 The Board's 28 executive members are divided into those from poor country borrowers and those from rich country donors. The latter have the most votes. Voting power is based upon cumulative contributions; hence the United States retains its majority share of the votes despite the fact that Japanese contributions are higher. For example, Japan provides 20 per cent of IDA replenishments and is only entitled to 10 per cent of the IDA vote. This bias was possibly a factor in Japan's decision to reduce its contribution to multilateral development institutions in 1997.

22 See Kenneth Piddington, 'The role of the World Bank', in Andrew Hurrell and Benedict Kingsbury (eds), *The International Politics of the Environment: Actors, Interests, and Institutions*, Oxford: Clarendon Press, 1992; Korinna Horta, 'The World Bank and the International Monetary Fund', in Jacob Werksman (ed.), *Greening International Institutions*, London: Earthscan, 1996; Robert Wade, 'Greening the Bank: the struggle over the environment, 1970–1995', in Kapur *et al.* (eds), *The World Bank*, volume 2; and Caufield, *Masters of Illusion*.

23 Wade, 'Greening the Bank', p. 724.

24 It should be noted here that despite its high international profile, the Bank's environment office was severely short staffed and largely marginalized within the institution at large. According to Bruce Rich, in 1983 the office still had only six staff (out of a total Bank staff of 6,000) with the impossible task of evaluating over 250 loan projects per annum. Rich, *Mortgaging the Earth*, p. 111.

25 Much has been written on the issue of World Bank-funded environmental disasters mostly from the NGO perspective. See Stephan Schwartzman, *Bankrolling Disasters: International Banks and the Global Environment*, San Francisco, CA: Sierra Club, 1986; and Graham Searle, *Major World Bank Projects: Their Impact on People, Society and the Environment*, Canelford, Cornwall: Wadebridge Ecological Centre, 1987. Numerous studies from the field date back to the early 1970s, see for example Mitaghi Farrar and John Milton, *The Careless Technology: Ecology and International Development*, Garden City, NY: Natural History Press, 1972. During the 1980s, when the environmental movement against the Bank was at its height, many field studies were also written by developing country NGOs. See Claude Alvares and Ramesh Billarey, *Damming the Narmada*, Penang: Third World Network and Asia Pacific People's Environment Network, 1988.

26 Wade, 'Greening the Bank'.

27 Rich, *Mortgaging the Earth*, p. 28.

28 Wade maintains that the impetus for environmental reform at the Bank was not simply a response to the environmental movement and like-minded members of the US Congress. He argues that a key motivating factor was the US Treasury, which under the leadership of James Baker made a tactical decision to support environmental sympathizers in Congress. This was to ensure Congressional support for a US capital contribution to IBRD in order to provide funds to offset the Latin American debt crisis. Wade, 'Greening the Bank', pp. 667–8.

29 Ibid., p. 655.

30 For example, Korinna Horta argues that in relation to forestry projects natural capital is highly likely to be further depleted if the foreign exchange is not available to pay back the loan. Horta, 'The World Bank and the International Monetary Fund', p. 135.

31 Charles Feinstein, Odil Payton and Kerri Poore, 'Global climate change – facing up to the challenge of Kyoto', in *Environment Matters at the World Bank: Annual Review*, Washington, DC: World Bank, Fall 1998.

32 Ibid., p. 51.

33 This appears to be a useful strategic alliance. The WWF has extensive experience in conservation and community forestry management but to succeed it needs the cooperation of the timber industry. The Bank's role is to persuade the timber industry to cooperate. Several forestry activities are currently underway.

34 Ken Newcombe, Juergen Blaser and Kerstin Canby, 'The World Bank and forests', in *Environment Matters*, Fall 1998.

35 World Bank, *Making Sustainable Commitments: An Environment Strategy for the World Bank*, Washington, DC: World Bank, 2001.

36 Urban environmental management projects largely focus upon water supply, sewage collection and treatment, solid waste management, traffic control and institutional development.

37 World Bank, 'Environmental projects portfolio', in *Environment Matters*, Fall 1998.

38 World Bank, 'The Bank's environment portfolio', in *Environment Matters at the World Bank*, Washington, DC: World Bank, June 2001.

39 World Bank, *World Bank Annual Report on the Environment*, Washington, DC: World Bank, 1991.

40 World Bank, *Making Sustainable Commitments*, p. 24.

41 World Bank, *Focus on Sustainability 2004*, Washington, DC: International Bank for Reconstruction and Development/World Bank, January 2005, p. 53.

42 The core function of the Environment Department is to work on environmental quality assessments to be built into the Bank's lending projects in both the rural and urban sectors. It does not have its own lending programme, although it does have some influence over the priorities of the regional divisions and country departments.

43 See World Bank, *Making Sustainable Commitments*.

44 Michael Redclift, 'Sustainable development and popular participation: a framework for analysis', in Dharam Ghai and Jessica Vivian (eds), *Grassroots Environmental Action: People's Participation in Sustainable Development*, New York: Routledge, 1992.

45 A Bank study carried out in 1992, for example, revealed that in Latin American cities water bought from vendors cost between 4 and 100 times the cost of piped water. More telling was the fact that in Lima a poor family was using roughly one-sixth as much water as a rich family but its water bills were three times more expensive. 'A greener bank', *Economist*, 23 May 1992. For a good overview of the redistributive effects of appropriate pricing policies, see Ismail Serageldin, *Water Supply, Sanitation, and Environmental Sustainability: The Financing Challenge*, World Bank Directions in Development Series, Washington, DC: World Bank, 1994.

46 Piddington, 'The role of the World Bank'.

47 World Bank, *World Bank Development Report: Development and the Environment*, Washington, DC: World Bank, 1992, p. 178.
48 Wade, 'Greening the Bank', p. 713.
49 'Let them eat pollution', *Economist*, 8 February 1992, p. 62. The memorandum is clearly provocative and, with the exception of hardline neo-classical economists, deeply offensive to most ears. Yet, the concluding remark in the memorandum (often ignored by the media at the time) not only puts the earlier thoughts into context but also cuts to the heart of the development problematique at the World Bank. It states:

> The problem with the arguments against all of these proposals for more pollution in LDCs (intrinsic rights to certain goods, moral reasons, social concerns, lack of adequate markets, etc.) could be turned around and used more or less effectively against every Bank proposal for liberalisation.

50 See John Dixon, *Environmental Economics and the Bank*, Washington, DC: World Bank, 1994; and Shakeb Afsah, Benoît Laplante and David Wheeler, *Controlling Industrial Pollution: A New Paradigm*, Washington, DC: World Bank, 1996.
51 Major Bank studies on the impact of pollution upon health and economic productivity include World Bank, *Can the Environment Wait?*, Washington, DC: World Bank, 1997; World Bank, *Clear Water and Blue Skies: China's Environment in the Twenty-First Century*, Washington, DC: World Bank, 1997; World Bank, *Managing Pollution Problems*, Washington, DC: World Bank, 1998; and World Bank, *Transition Towards a Healthier Environment: Environmental Issues and Challenges in the Newly Industrialised States*, Washington, DC: World Bank, 1998.
52 World Bank, Environmentally Sustainable Development, *The World Bank Participation Sourcebook*, Washington, DC: World Bank, 1996.
53 Devesh Kapur, John Lewis and Richard Webb (eds), *The World Bank: Its First Half Century*, volumes 1 and 2, Washington, DC: Brookings Institution, 1997.
54 World Bank, China, *The World Bank Group in China, Facts and Figures*, Beijing: World Bank, 2002.
55 World Bank internal memorandum (confidential), 9 April 1999.
56 World Bank, China, *The World Bank Group in China, Facts and Figures*, Beijing: World Bank, 2003.
57 World Bank, *World Bank Development Report: China Urban Environmental Service Management*, Report No. 13073-CHA, Washington, DC: World Bank, 1994, p. xiii.
58 World Bank, *Clear Water and Blue Skies*, pp. 2–3.
59 Other environmental sector reports have been written but these are more specific such as World Bank, *China: Efficiency and Environmental Impact of Coal*, Washington, DC: World Bank, 1991; and World Bank, *World Bank Development Report: China: Involuntary Resettlement*, Washington, DC: World Bank, 1993.
60 This will be explored in detail towards the end of this chapter.
61 The Bank's East Asia Sustainable Environment Network, which is part of the Environment Department, has been involved in the technical assistance programme to SEPA.
62 World Bank, *China, Air, Land, and Water: Environmental Priorities for a New Millennium*, Washington, DC: World Bank, 2001.
63 Calculated from World Bank, 'Environmental projects portfolio'.
64 Interview, East Asia Sustainable Environment Network, Environment Department, World Bank, Washington, DC, 9 April 1999.
65 Horta, 'The World Bank and the International Monetary Fund', p. 135.
66 World Bank, *Environment Matters at the World Bank: Annual Review*, Washington, DC: World Bank, July 2002–June 2003, p. 43.
67 World Bank, *China 2020: Development Challenges in the New Century*, Washington, DC: World Bank, 1997.

68 Devesh Kapur, John Lewis and Richard Webb, 'Introduction', in Kapur, Lewis and Webb (eds), *The World Bank*, volume 1, p. 25.
69 Nick Young, 'China always in the driving seat', *China Development Briefing*, 1(4), 1997, p. 5.
70 Interview, Urban Development Sector Unit, East Asia and Pacific Region, World Bank, Washington, DC, 6 April 1999.
71 For example, the 2001 report on China's environment was prepared jointly by the World Bank and a team of specialists from ten Chinese research institutes, universities and NGOs.
72 Interview, World Bank Resident Mission, Beijing, 1 July 1998.
73 For example, the Bank's 1998 project on reducing ozone depletion in China aims to reduce the production and consumption of halons by introducing a tradable production quota and bidding system. See www.worldbank.org/ (accessed 30 May 2000).
74 See World Bank, *Clean Development Mechanism in China: Taking a Proactive and Sustainable Approach*, Washington, DC: International Bank for Reconstruction and Development/The World Bank, June 2004.
75 The withdrawal of IDA funding was based on a decision made by the Bank's Board of Directors to the effect that China should no longer be classified as a low-income developing country.
76 Interview, Foreign Cooperation Department, SEPA, Beijing, 28 April 1999.
77 Ironically, the Bank's last soft loan project to China – the Western Poverty Reduction project – was withdrawn. This project aimed to relocate 58,000 people from the eroded uplands of Haidong prefecture in Qinghai to the Tibetan plains. The communities involved cannot survive – unable to sustain even subsistence farming because of increasing ecological degradation. But the project stimulated strong opposition from international NGOs and politicians in the United States, who believed that the project was merely a cover-up for the 'Han colonization' of Tibet. In fact, over 50 per cent of the Qinghai population that were expecting to relocate to Tibet were of Hui or Salar ethnic origin. See Nick Young, ' "Ethnic diluting" row clouds last soft bank loan', *China Development Brief*, 2(3), 1999, www.chinadevelopmentbrief.com/article. asp?sec=19&sub=1&toc=11&art=150. With the consent of the Chinese government the Bank set up an inspection panel which submitted its findings to the Bank's Board of Directors. The Board requested further review and approval before the original project recommendations could be accepted. The Chinese authorities have since decided to finance the project with their own resources and have been quick to point out the political nature of the Bank's decision-making authority. World Bank website www.worldbank.org/news/pres (accessed 10 July 2000).
78 Interviews, Urban Development Sector Unit, World Bank, Washington, DC, 6 April 1999; and Industry Department, World Bank, Washington, DC, 9 April 1999.
79 Interview, World Bank Resident Mission, Beijing, 1 July 1998.
80 FBIS Beijing *Xinhua*, 9 September 1997.
81 *China Daily*, 25 June 1998.
82 Interview, Energy Division, World Bank, Washington, DC, 8 April 1999.
83 Water supply companies in China have a long history. Many were established before 1949. For the most part, these companies remain government owned and managed. But utilities for wastewater and solid waste management are few and far between. The Bank's objective, therefore, is to create utilities specifically to cater for these services and, in so doing, ensure that they are commercially viable by promoting financial autonomy from local government.
84 World Bank Staff Appraisal Report No. 12708-CHA, *China Liaoning Environment Project*, Washington, DC: Urban Development Sector Unit, East Asia and Pacific Regional Office, July 1994, p. 13.
85 Heritage conservation is an interesting addition to the Bank's portfolio of urban environmental initiatives and is still at the early stages of development. However, the

concept appeared to be a peripheral concern for local project officials in Shenyang, which may have something to do with the fact that tourism is under-developed in Liaoning, and hence the economic incentive does not exist to conserve cultural assets. Even when the political will does exist, one of the problems with cultural conservation in general is that it is not always clear who should take responsibility, nor, often, what constitutes 'cultural' or how to conserve it. World Bank internal memorandum (confidential), 26 February 1999.

86 World Bank Staff Appraisal Report No. 12708-CHA, *China Liaoning Environment Project.*

87 Even if local governments are committed to raising water prices, the price increase still has to be approved by the Central Pricing Bureau in Beijing. Interview, World Bank Resident Mission, Beijing, 28 April 1999.

88 Interview, European Union Project Office, Liaoning EPB, 11 May 1999.

89 Industry was paying approximately 10–20 per cent more for its wastewater treatment depending upon whether it involved primary treatment ($RMB1.2/m^3$) or biological treatment ($RMB0.5/m^3$). In addition, household tariffs differed between cities. At $RMB2.5/m^3$, charges in Dalian were higher than those in Shenyang. Interview, LUCRPO, 13 May 1999.

90 Interview, World Bank Resident Mission, Beijing, 1 July 1998.

91 Ibid.

92 Interviews, Jinzhou World Bank sub-project office, 20 May 1999; and Dalian World Bank sub-project office, 24 May 1999.

93 Interview, LUCRPO, Shenyang, 13 May 1999.

94 Project site visit to Jincheng Paper Mill, Jinzhou, 23 May 1999.

95 Interview, European Union Project Office, Liaoning EPB, 11 May 1999.

96 Interview, LUCRPO, 13 May 1999.

97 Yunnan Environmental Protection Bureau, *Yunnan sheng huanjing baohu jiuwu jihua he 2010 nian chang yuan guihua* (Yunnan Province Environmental Protection Ninth Five-Year Plan and Long Term Plan until 2010), Kunming: Yunnan EPB, 1996.

98 Ibid.

99 Interviews, Yunnan EPB, 10 July 1998, and 8 June 1999.

100 Yunnan Environmental Protection Bureau, *Yunnan sheng shijie yinhang daikuan huanbao xiangmu jieshao* (Introduction to the Yunnan World Bank Environmental Loan Project), Kunming: Yunnan EPB, 1996.

101 Eutrophication typically occurs when phosphorous concentrations are around 0.2 µg/l. By 1996 a report showed that phosphorous concentrations in the lake were as high as 0.55 µg/l. Yunnan Institute of Environmental Sciences, *Environmental Assessment Report*, Kunming: Yunnan EPB, February 1996.

102 Interview, Yunnan EPB, 8 June 1999.

103 Interview, Rural Development Research Centre, Yunnan Institute of Geography, 11 July 1998.

104 Interview, Yunnan World Bank Project Office, 8 June 1999.

105 Wu Jiachun, 'Yunnan battles to save the Dianchi Lake', *China Daily*, 29 October 1998.

106 At least 50 per cent of wastewater discharges flow to the small inner lake (Caohai) which accounts for only 2.5 per cent of the total lake area. Consequently, the dyke was constructed to cut off the Caohai from the outer lake (Waihai) and thereby control the flow of pollution.

107 Phosphorous loads were identified as predominantly point source outputs from domestic and industrial wastewater, although the local scientific report carried out in 1995 also revealed the importance of non-point source outputs such as fertilizer run-offs and village wastes. Out of a total of 781 tons of phosphorous dumped into the lake every year, approximately 479 tons originate from point sources and 302 tons from non-point sources. In the Bank's Staff Appraisal Report in 1996 the significance of

the latter was downplayed with the suggestion that the data for non-point sources was based upon limited measurements. Yunnan Institute of Environmental Services, *Environmental Assessment Report*; and World Bank Staff Appraisal Report No. 15361-CHA, *China Yunnan Environment Project*, Washington, DC: Urban Development Sector Unit, East Asia and Pacific Regional Office, May 1996.

108 World Bank Staff Appraisal Report No. 15361-CHA, *China Yunnan Environment Project*, p. 15.

109 Interview, Yunnan World Bank Project Office, 8 June 1999.

110 Prior to the tariff adjustment the Bank had conducted two willingness-to-pay surveys in Kunming. A total of 73 per cent of respondents said that they would be willing to pay up to RMB0.60/m³ for their water supply. World Bank Staff Appraisal Report No. 15361-CHA, *China Yunnan Environment Project*, p. 188.

111 Interview, Yunnan World Bank Project Office, 8 June 1999.

112 World Bank Staff Appraisal Report No. 15361-CHA, *China Yunnan Environment Project*.

113 Interview, Yunnan World Bank Project Office, 8 June 1999.

114 Interview, Yunnan EPB, 10 July 1998.

115 Interview, Yunnan World Bank Project Office, 8 June 1999.

116 Ibid.

117 The Water Research Institute at the CRAES in Beijing had been asked by the Bank to provide additional institutional support but according to the Soil and Fertilizer Station, this was largely inadequate. CRAES did not have the capacity to carry out both the research and design requirements for the project. Furthermore, it was deemed to be out of touch with the needs of the villagers. Interview, Soil and Fertilizer Station, Kunming Agricultural Bureau, 9 June 1999.

118 By attracting surplus labour from the rural areas, the growth of TVEs has been a major factor behind the transformation of rural villages into semi-urban settlements.

119 Interview, Rural Development Research Centre, Yunnan Institute of Geography, 11 July 1998.

120 Of course the Bank is not alone in making this distinction. Most other multilateral and bilateral donors follow suit.

121 World Bank Staff Appraisal Report No. 15361-CHA, *China Yunnan Environment Project*, p. 52.

122 During the design stage, it was still too early for the Bank to have drawn any lessons from its experience in Yunnan, but it had accumulated a number of lessons from working in the poorer areas of other Chinese cities such as Tianjin.

123 It is important to stress here that, compared with other cities in China, the municipal governments in Guangxi are more tolerant of temporary residents. They have provided some basic services such as schooling and garbage collection but wastewater and sewage control have largely been ignored. World Bank Staff Appraisal Report No. 16622-CHA, *China Guangxi Urban Environment Project*, Washington, DC: Urban Development Sector Unit, East Asia and Pacific Regional Office, May 1998.

124 This component involves the extension of a 26-km canal to convey water from a reservoir to the river, the installation of a 0.8 MW hydropower generator, and the reparation of certain sections of the river's embankment.

125 World Bank Staff Appraisal Report No. 16622-CHA, *China Guangxi Urban Environment Project*.

126 In 1998, industrial tariffs were also increased to RMB0.64/m³ for water (including RMB0.18 for wastewater) in Nanning and RMB0.88/m³ for water (including RMB0.20 for wastewater) in Guilin. Interview, GUEPO, Guangxi Regional EPB, 17 June 1999.

127 In Guilin a sewerage system has been in operation since the early 1970s, which reflects the city's historic concern with maintaining water quality for the purposes of tourism.

128 World Bank Staff Appraisal Report No. 16622-CHA, *China Guangxi Urban Environment Project*, p. 22.
129 Interview, GUEPO, Guangxi Regional EPB, 17 June 1999.
130 Interview, Urban Development Sector Unit, East Asia and Pacific Region, World Bank, Washington, DC, 8 April 1999.
131 The World Bank–UNDP Water and Sanitation Program was initiated in 1978 in the lead up to the United Nations Drinking Water Supply and Sanitation Decade 1981–1990. The program focuses upon the provision of 'hardware' and 'software' in relation to water supply and sanitation in the world's poorest communities. It is managed within the World Bank and only has a small budget of approximately US$15 million per annum. However, its influence upon World Bank policy and practice has been considerable in proportion to its resources. For more detailed information on the demand-responsiveness approach see Maggie Black, *Learning What Works: A 20 Year Retrospective View on International Water and Sanitation Cooperation 1978–1998*, Washington, DC: UNDP–World Bank Water and Sanitation Program, September 1998.
132 Interview, Nanning World Bank Project Office, 18 June 1999.
133 Interview, Yangshuo World Bank Project Office, 26 June 1999.
134 Interview, Nanning World Bank Project Office, 18 June 1999.
135 Ibid.
136 Ibid.
137 Interview, Guilin World Bank Project Office, 25 June 1999.
138 Interview, Yangshuo World Bank Project Office, 26 June 1999.
139 Interview, Urban Development Sector Unit, East Asia and Pacific Region, World Bank, Washington, DC, 8 April 1999.

6 The promises and pitfalls of international environmental aid to China

1 See Martin Jänicke and Hans Weidner (eds), *National Environmental Policies: A Comparative Study of Capacity-Building*, Berlin: Springer, 1997; Susan Baker and Petr Jehlicka, 'Dilemmas of transition: the environment, democracy and economic reform in east central Europe – an introduction', *Environmental Politics*, 7(1), 1998; and Elizabeth Wilson, 'Capacity for environmental action in Slovakia', *Journal of Environmental Planning and Management*, 42(4), 1999.
2 Tom H. Tietenberg, 'Economic instruments for environmental regulation', *Oxford Review of Economic Policy*, 6(1), 1990; Jänicke and Weidner (eds), *National Environmental Policies*.
3 In China, staff shortages within government agencies are not generally a problem. The EPBs are no exception. Recall that the number of personnel working at the EPBs in Shenyang and Kunming, for example, did not differ considerably. More important are the educational background of staff and the quality of their skills.
4 *Japan Times*, 19 August 2002.
5 *Japan Times*, 5 September 2002.
6 Hideaki Shiroyama, 'Clean and efficient coal use in China and the political economy of international aid', *Social Science Japan*, 16, August 1999.
7 Internal memorandum (confidential), Urban Development Sector Unit, World Bank, Washington, DC, 9 April 1999.
8 Kenji Yoshida, Managing Director, Operations Department II, OECF, Tokyo. Cited in OECF Newsletter, *Toward Improvement of the Environment and Alleviation of Poverty*, Tokyo: OECF, March 1999.
9 Confidential source.
10 Interview, Planning Department, JICA, Tokyo, 26 March 1999.
11 Japan Aid Study Group on Environment, *Japan International Cooperation Agency Sectoral Study for Development Assistance*, Tokyo: JICA, December 1988, p. 44.

12 Motomichi Ikawa, 'JBIC: a new force in global development', *Look Japan*, 'Initiatives for sustainable development: Japan's environmental ODA programs', 45(5), 24 November, 1999.
13 www.jbic.go.jp (accessed 10 February 2000).
14 Interview, Social Development and Poverty Alleviation Division, UNDP, New York, 31 March 1999.
15 Interview, GEF regional coordinator for climate change, UNDP, New York, 1 April 1999.
16 Interview, UNIDO, Beijing, 2 July 1998.
17 I am grateful to Ai Chin Wee from the World Bank for drawing my attention to this point.
18 Interview, Urban Development Sector Unit, East Asia and Pacific Region, World Bank, Washington, DC, 8 April 1999.
19 Interview, Habitat advisor, Wuhan EPB, 2 July 1999.
20 F. Danida, *Effectiveness of Multilateral Agencies at Country Level: Case Study of 11 Agencies in Kenya, Nepal, Sudan, and Thailand*, Copenhagen: Ministry of Foreign Affairs, 1991.

Conclusion

1 At the 2002 World Summit on Sustainable Development, the international community made a commitment to strengthen their response to global poverty and environmental decline. For an overview of the new policy thinking in this regard see Department for International Development, Directorate General for Development, European Commission, UNDP and the World Bank, 'Linking poverty reduction and environmental management: policy challenges and opportunities', Working Paper, Washington, DC: World Bank, July 2002.
2 See World Bank, 'Memorandum of the president of IBRD and IFC to the executive directors in a Country Assistance Strategy of the World Bank Group for the People's Republic of China', Washington, DC: China Country Management Unit, East Asia and Pacific Region, 22 January 2003.
3 www.worldbank.org.cn (accessed 8 May 2003).
4 'JBIC extends ¥161,366 million in ODA loans to China – to support environment protection and human resources and regional development', 29 March 2002, www.jbic.go.jp/autocontents/english/news/2002/000019/index.htm (accessed 3 March 2002).
5 Jonathan Unger, *The Transformation of Rural China*, Armonk, NY: M.E. Sharpe, 2002, p. 178.
6 See Paul Mosley, Jane Harrigan and John Toye, *Aid and Power: The World Bank and Policy-Based Lending*, volumes 1 and 2, London: Routledge, 1991; and Joan M. Nelson and Stephanie J. Eglinton, *Global Goals, Contentious Means: Issues of Multiple Aid Conditionality*, Washington, DC: Overseas Development Council, 1993.
7 Alex Duncan, 'Aid effectiveness in raising adaptive capacity in the low-income countries', in John Lewis and Valerina Kallab (eds.), *Development Strategies Reconsidered*, Washington, DC: Overseas Development Council, 1985; and Mosley, Harrigan and Toye, *Aid and Power*, 1991.

Epilogue: can lessons be learned?

1 Louise Lim, 'China warns of water pollution', BBC news report, Beijing, 23 March 2005, http://news.bbc.co.uk/2/hi/asia-pacific/4374383.stm (accessed 27 March 2005).
2 Andreas Lorenz, 'China's environmental suicide: a government minister speaks', interview with Pan Yue, 5 April 2005, www.opendemocracy.net/debates/article-6–129–2407.jsp (accessed 8 April 2005).
3 *Renmin Ribao*, 17 November 2004.

4 www.zhb.gov.cn (accessed 15 March 2005).
5 For further information see the campaign website www.nujiang.ngo.cn
6 *Economist*, 18 October 2003, p. 11.
7 *Yomiuri Shimbun*, 18 March 2005.
8 I would like to thank officials from the UNDP, Japan Bank for International Cooperation, and the World Bank for keeping me up to date with project implementation, and for making available the evaluations that have been completed.
9 World Bank, *Implementation Completion Report on a Loan in the Amount of US$110 Million to the People's Republic of China for a Liaoning Environment Project*, Report No. 28833-CHA, Washington, DC: Urban Development Sector Unit, East Asia and Pacific Region, 22 June 2004, p. 41.
10 *Renmin Ribao*, 15 December 2003, cited in Nick Young, 'Bilateral donors trim budgets, advance governance and policy agendas', *China Development Brief*, 7(1), Spring 2004.

Bibliography

Afsah, Shakeb, Benoît Laplante and David Wheeler, *Controlling Industrial Pollution: A New Paradigm*, Washington, DC: World Bank, 1996.

Agrawal, Arun, 'Dismantling the divide between indigenous and scientific knowledge', *Development and Change*, 26(3), 1995, pp. 413–39.

Alvares, Claude and Ramesh Billarey, *Damming the Narmada*, Penang: Third World Network and Asia Pacific People's Environment Network, 1988.

Amsden, Alice H., Dongyi Liu and Xiaoming Zhang, 'China's macroeconomy, environment, and alternative transition model', *World Development*, 24(2), 1996, pp. 273–86.

Anderson, Terry and Donald Leal, *Free Market Environmentalism*, Boulder, CO: Westview, 1991.

Arase, David, 'Public–private sector interest coordination in Japan's ODA', *Pacific Affairs*, 67(2), 1994, pp. 171–99.

——, *Buying Power: The Political Economy of Japan's Foreign Aid*, Boulder, CO: Lynne Rienner, 1995.

Arndt, Heinz W., 'Sustainable development and the discount rate', in Heinz W. Arndt, *50 Years of Development Studies*, Canberra: National Centre for Development Studies, Australian National University, 1993.

Arrow, Kenneth, Bert Bolin, Robert Costanza, Partha Dasgupta, Carl Folke, C.S. Holling, Bengt-Owe Jansson, Simon Levin, Karl-Goran Maler, Charles Perrings and David Pimental, 'Economic growth, carrying capacity, and the environment', *Science*, 268(5210), 1995, pp. 520–1.

Baker, Susan and Petr Jehlicka, 'Dilemmas of transition: the environment, democracy and economic reform in east central Europe – an introduction', *Environmental Politics*, 7(1), 1998, pp. 1–26.

Banister, Judith, 'Population, public health and the environment in China', *China Quarterly*, 156(December), 1998, pp. 986–1015.

Beck, Ulrich, *The Risk Society: Towards a New Modernity*, trans. Mark Ritter, London: Sage, 1992.

Beck, Ulrich, Anthony Giddens and Scott Lash, *Reflexive Modernization: Politics, Tradition and Aesthetics in the Modern Social Order*, Cambridge: Polity Press in association with Blackwell Publishers, 1994.

Benxi's Agenda 21 Leading Group Office, 'Benxi Agenda 21: overall design for the sustainable development strategy of Benxi City', unpublished, Benxi, 1994.

Binder, Leonard, James S. Coleman, Joseph LaPalombara and Lucian W. Pye, *Crises and Sequences in Political Development*, Princeton, NJ: Princeton University Press, 1971.

Black, Maggie, *Learning What Works: A 20 Year Retrospective View on International Water and Sanitation Cooperation 1978–1998*, Washington, DC: UNDP–World Bank Water and Sanitation Program, September 1998.

Bohm, Robert A., Chazhong Ge, Milton Russell, Jinnan Wang and Jintian Yang, 'Environmental taxes: China's bold initiative', *Environmental Politics*, 40(7), 1998, pp. 10–14.

Breslin, Shaun, *China in the 1980s: Centre–Province Relations in a Reforming Socialist State*, London: Macmillan, 1996.

——, 'China's environmental crisis in a global context', *Global Society: Journal of Interdisciplinary International Relations*, 10(2), 1996, pp. 125–44.

Cadieux, Louise, 'Liaoning in transition of reform: what will become of the state-owned enterprises', *China Today*, 48(11), 1999, pp. 10–13.

Carruthers, Ian and Roy Stoner, 'Economic aspects and policy issues in ground water development', World Bank Staff Working Paper No. 496, Washington, DC: World Bank, 1981.

Cassen, Robert, *Does Aid Work?*, 2nd edn, Oxford: Clarendon Press, 1994.

Cassen, Robert and Associates, *Does Aid Work? Report to an Intergovernmental Task Force*, Oxford: Oxford University Press, 1986.

Caufield, Catherine, *Masters of Illusion: The World Bank and the Poverty of the Nations*, London: Pan Books, 1998.

Chambers, Robert and Gordon Conway, 'Sustainable rural livelihoods: practical concepts for the twenty-first century', IDS Discussion Paper 296, Brighton: Institute of Development Studies, 1992.

Chan, Anita, 'Revolution or corporatism? Private entrepreneurs as citizens: from Leninism to corporatism', *China Information*, 10(3/4), 1995–1996, pp. 1–28.

Chan, Hon, Koon-Kwai Wong, K.C. Cheung and Jack Man-Keung Lo, 'The implementation gap in environmental management in China: the case of Guangzhou, Zhengzhou, and Nanjing', *Public Administration Review*, 55(4), 1995, pp. 333–40.

Chan, Hon, K.C. Cheung and Jack M.K. Lo, 'Environmental control in the PRC', in Stuart Nagel and Miriam Mills (eds), *Public Policy in China*, Westport, CT: Greenwood Press, 1993, pp. 63–81.

Chan, Kam Wing, *Cities with Invisible Walls: Reinterpreting Urbanization in Post-1949 China*, Hong Kong: Oxford University Press, 1994.

Chatterjee, Pertap and Matthias Finger, *The Earth Brokers: Power, Politics, and World Development*, London and New York: Routledge, 1994.

Cheung, Peter and Jae Ho Chung (eds), *Provincial Strategies of Economic Reform in Post-Mao China: Leadership, Politics and Implementation*, Armonk, NY: M.E. Sharpe, 1998.

China National Environmental Protection Agency, *China's TransCentury Green Plan 1996–2000*, Beijing: Zhongguo huanjing kexue chubanshe, 1997.

CICETE (China International Center for Economic and Technical Exchanges), *China/UNDP The Fourth China Country Programme: Environment and Energy Projects*, Beijing: CICETE, 1996.

——, *UNDP Air Pollution Control Programme*, Beijing: CICETE, 1996.

Clarke, John, *Democratizing Development: The Role of Voluntary Organization*, London: Earthscan, 1991.

Cohen, John, 'Foreign advisors and capacity building: the case of Kenya', *Public Administration and Development*, 12(5), 1993, pp. 493–510.

Commoner, Barry, *The Closing Circle: Confronting the Environmental Crisis*, London: Cape, 1972.

Dai Qing, *Yangtze! Yangtze!*, London: Earthscan, 1994.

Dalian Environmental Protection Bureau, 'Promotion of the construction plan of Dalian environmental demonstration zone: towards sustainable development', unpublished, Dalian: Dalian EPB, 1998.

Daly, Herman (ed.), *Steady-State Economics*, San Francisco, CA: W.H. Freeman, 1977.

Danaher, Kevin (ed.), *50 Years is Enough: The Case Against the World Bank and the International Monetary Fund*, Boston, MA: Southend Press, 1994.

'Dangdai Zhongguo' cong shu bian ji bu (ed.), *Dangdai zhongguo de duiwai jingji hezuo* (Contemporary Chinese Economic Cooperation with Foreign Countries), Beijing: Zhongguo shehui kexue chubanshe, 1989.

Danida, F., *Effectiveness of Multilateral Agencies at Country Level: Case Study of 11 Agencies in Kenya, Nepal, Sudan, and Thailand*, Copenhagen: Ministry of Foreign Affairs, 1991.

Danish, Klye, 'The promise of national environmental funds in developing countries', *International Environmental Affairs*, 7(2), 1995, pp. 150–75.

Dauvergne, Peter, *Shadows in the Forest: Japan and the Politics of Timber in Southeast Asia*, Cambridge, MA: MIT Press, 1997.

Delfs, Robert, 'Poison in the sky', *Far Eastern Economic Review*, 156(5), 4 February 1993, p. 16.

Department for International Development, Directorate General for Development, European Commission, United Nations Development Programme and the World Bank, 'Linking poverty reduction and environmental management: policy challenges and opportunities', Working Paper, Washington, DC: World Bank, July 2002.

Ding, Yijiang, 'Corporatism and civil society in China: an overview of the debate in recent years', *China Information*, 12(4), 1998, pp. 44–67.

Dixon, John, *Environmental Economics and the Bank*, Washington, DC: World Bank, 1994.

Douglas, Amy, *The Politics of Environmental Mediation*, New York: Columbia University Press, 1987.

Dower, John, *Embracing Defeat: Japan in the Aftermath of World War II*, London: Allen Lane, 1999.

Dryzek, John, *The Politics of the Earth: Environmental Discourses*, Oxford: Oxford University Press, 1997.

Duncan, Alex, 'Aid effectiveness in raising adaptive capacity in the low-income countries', in John Lewis and Valeriana Kallab (eds), *Development Strategies Reconsidered*, Washington, DC: Overseas Development Council, 1985, pp. 129–52.

Durkheim, Emile, [On the] *Division of Labour in Society*, trans. G. Simpson, New York: Macmillan, 1933, first published 1911.

Eade, Deborah, *Capacity-Building: An Approach to People-Centred Development*, Oxford: Oxfam (UK and Ireland), 1997.

Eaton, Joseph (ed.), *Institution Building and Development: From Concepts to Application*, Beverly Hills, CA: Sage Publications, 1972.

Eckersley, Robyn (ed.), *Markets, the State and the Environment: Towards Integration*, South Melbourne: Macmillan Education Australia, 1995.

Economy, Elizabeth C., 'The case of China', Environmental Scarcities, State Capacity, and Civil Violence Project, CISS Occasional Paper, Cambridge, MA: Committee on International Security Studies, American Academy of Arts and Sciences, 1997.

——, *The River Runs Black: The Environmental Challenge to China's Future*, Ithaca, NY and London: Cornell University Press, 2004.

Edmonds, Richard Louis, *Patterns of China's Lost Harmony: A Survey of the Country's Environmental Degradation and Protection*, London: Routledge, 1994.

—— (ed.), *Managing the Chinese Environment*, New York: Oxford University Press, 1998.

Ekins, Paul, *A New World Order: Grassroots Movements For Global Change*, London: Routledge, 1992.

Elvin, Mark, 'The environmental legacy of imperial China', *China Quarterly*, 156(December), 1998, pp. 757–87.

Elvin, Mark and Liu Ts'ui-jung, *Sediments of Time: Environment and Society in Chinese History*, Cambridge: Cambridge University Press, 1998.

Engberg-Pedersen, Poul and Claus Hvashøj Jørgensen, 'UNDP and global environmental problems: the need for capacity development at country level', in Helge Ole Bergesen and Georg Parmann (eds), *Green Globe Yearbook of International Co-operation on Environment and Development*, Oxford: Oxford University Press, 1997, pp. 37–43.

Esman, Milton, 'Institution building in national development', in Gove Hambridge (ed.), *Dynamics of Development*, New York: Praeger, 1964, pp. 140–51.

——, *Administration and Development in Malaysia: Institution Building and Reform in a Plural Society*, Ithaca, NY: Cornell University Press, 1972.

——, *Management Dimensions of Development: Perspectives and Strategies*, West Hartford, CT: Kumarian Press, 1991.

Eto, Shinkichi, 'China and Sino-Japanese relations in the coming decades', *Japan Review of International Affairs*, 10(1), 1996, pp. 16–34.

Evans, Peter, 'Japan's green aid', *China Business Review*, July–August, 1994.

Fairman, David and Michael Ross, 'Old fads, new lessons: learning from economic development assistance', in Robert Keohane and Marc Levy (eds), *Institutions for Environmental Aid: Pitfalls and Promise*, Cambridge, MA: MIT Press, 1996, pp. 29–51.

Falk, Richard, 'The global promise of social movements: explorations at the edge of time', *Alternatives*, 12(2), 1987, pp. 173–96.

Farrar, Mitaghi and John Milton, *The Careless Technology: Ecology and International Development*, Garden City, NY: Natural History Press, 1972.

Feinerman, James, 'Chinese participation in the international legal order: rogue elephant or team player?', *China Quarterly*, 141(March), 1995, pp. 186–210.

Feinstein, Charles, Odil Payton and Kerri Poore, 'Global climate change – facing up to the challenge of Kyoto', in *Environment Matters at the World Bank: Annual Review*, Washington, DC: World Bank, Fall 1998.

Findlay, Christopher, Andrew Watson and Harry Wu (eds), *Rural Enterprises in China*, London: Macmillan, 1994.

Fisher, Duncan, *Paradise Deferred: Environmental Policymaking in Central and Eastern Europe*, London: Energy and Environment Programme, Royal Institute of International Affairs, 1992.

Gant, George F., 'The institution building project', *International Review of Administrative Sciences*, 32(3), 1966, pp. 219–25.

Gao, Dazhi, Wang Jinjia and Cai Jinlui, 'Mercury pollution and control in China', *Journal of Environmental Sciences (China)*, 3(3), 1991, pp. 105–21.

George, Susan and Fabrizio Sabelli, *Faith and Credit: The Bank's Secular Empire*, Washington, DC: Institute for Policy Studies, 1994.

Georgescu-Roegan, Nicholas, *The Entropy Law and the Economic Process*, Cambridge, MA: Harvard University Press, 1971.

Ghai, Dharam and Jessica Vivian (eds), *Grassroots Environmental Action: People's Participation in Sustainable Development*, London: Routledge, 1992.

Goldenberg, José and Thomas Johansson, *Energy as an Instrument for Socio-Economic Development*, New York: UNDP, 1995.

Goodin, Robert, 'Selling environmental indulgences', *Kyklos*, 47(4), 1994, pp. 573–95.

Goodman, David (ed.), *China's Provinces in Reform: Class, Community, and Political Culture*, London: Routledge, 1997.

Gouldson, Andrew and Joseph Murphy, *Regulatory Realities: The Implementation and Impact of Industrial Environmental Regulation*, London: Earthscan, 1998.

Grindle, Merilee and Mary Hildebrand, 'Building sustainable capacity in the public sector: what can be done?', *Public Administration and Development*, 15(5), 1995, pp. 441–63.

Guangxi Regional Project Office, 'Start-up workshop for small area improvement component', prepared by the World Bank in consultation with the Regional Project Management Office for Guangxi Environment Project, Nanning, 4–6 February 1999.

Guldin, Gregory Eliyn (ed.), *Farewell to Peasant China: Rural Urbanization and Social Change in the Late Twentieth Century*, Armonk, NY: M.E. Sharpe, 1997.

Gunderson, Adolf, *The Environmental Promise of Democratic Deliberation*, Madison, WI: University of Wisconsin Press, 1995.

Guojia huanbaoju zhengce faguisi (ed.), *Huanjing baohu fagui huibian* (Compendium of Environmental Protection Laws and Regulations), Beijing: Zhongguo huanjing kexue chubanshe, 1990.

Haas, Peter M., 'Introduction: epistemic communities and international policy coordination', *International Organization*, 46(1), 1992, pp. 1–35.

Haas, Peter M., Robert Keohane and Marc Levy, *Institutions for the Earth: Sources of Effective International Environmental Protection*, Cambridge, MA: MIT Press, 1993.

Habitat/UNEP, *SCP 1997 Meeting Report*, Nairobi: Habitat/UNEP, 1998.

——, *Sustainable Cities Programme: Approach and Implementation*, Nairobi: Habitat/UNEP, 1998.

——, *The SCP Process Activities: A Snapshot of What They Are and How They Are Implemented*, Nairobi: Habitat/UNEP, 1998.

Hajer, Maarten, *The Politics of Environmental Discourse: Ecological Modernization and the Policy Process*, Oxford: Clarendon Press, 1995.

Hardin, Garrett, 'The tragedy of the commons', *Science*, 162(3859), 1968, pp. 1243–8.

Hashimoto, Michio, 'Development of environmental policy and its institutional mechanisms of administration and finance', paper presented at International Workshop on Environmental Management for Local and Regional Development, sponsored by UNCRD and UNEP, Nagoya, Japan, 1985.

——, *Economic Development and the Environment: The Japanese Experience*, Tokyo: Ministry of Foreign Affairs, 1992.

——, 'Development and environmental problems', *Asian Economic Journal*, 8(1), 1994, pp. 115–45.

He Bochan, *China on the Edge: Crisis of Ecology and Development*, San Francisco, CA: China Books and Periodicals, 1991.

He, Baogang, *The Dual Role of Semi-Civil Society in Chinese Democracy*, Brighton: Institute of Development Studies, University of Sussex, 1993.

Hirschman, Albert O., *Shifting Involvements: Private Interest and Public Action*, Oxford: Oxford University Press, 1982.

Ho, Peter, 'Greening without conflict? Environmentalism, NGOs and civil society in China', *Development and Change*, 32(5), 2001, pp. 893–921.

Hook, Steven W. and Guang Zhang, 'Japan's aid policy since the Cold War: rhetoric and reality', *Asian Survey*, 38(11), 1998, pp. 1051–66.

Horta, Korinna, 'The World Bank and the International Monetary Fund', in Jacob Werksman (ed.), *Greening International Institutions*, London: Earthscan, 1996, pp. 131–47.

Hoshina Hideaki, 'Dairen-shi toshi kankyô moderu chika keikaku chosa' (Basic research plan on Dalian environmental model city project), unpublished draft report, Tokyo: JICA, 1999.

Huang, Philip C.C., ' "Public sphere"/"civil society" in China? The third realm between state and society', *Modern China*, 19(2), 1993, pp. 216–40.

Huang, Yiping, 'State-owned enterprise reform', in Ross Garnaut and Ligang Song (eds), *China: Twenty Years of Economic Reform*, Canberra: NCDS Asia Pacific Press, 1999, pp. 95–116.

Huber, Joseph, *Die Verlorene Unchuld der Okologie*, Frankfurt: Fischer Verlag, 1982.

Human Rights China, *The Emerging Non-Profit Sector: Seeking Independence in a Regulatory Cage*, Beijing: Human Rights China, 13 November 1998.

Huntington, Samuel, *Political Order in Changing Societies*, New Haven, CT: Yale University Press, 1968.

Hurrell, Andrew, 'A crisis of ecological viability? Global environmental change and the nation state', in John Dunn (ed.), *Contemporary Crisis of the Nation State?*, Oxford: Blackwell Publishers, 1995, pp. 146–65.

Hurrell, Andrew and Benedict Kingsbury (eds), *The International Politics of the Environment: Actors, Interests and Institutions*, Oxford: Clarendon Press, 1992.

Hyden, Goran, 'Governance and sustainable livelihoods: challenges and opportunities', paper presented at Workshop on Sustainable Livelihoods and Sustainable Development, University of Florida, 1–3 October 1998.

Ikawa, Motomichi, 'JBIC: a new force in global development', *Look Japan*, 'Initiatives for sustainable development: Japan's environmental ODA programs', 45(5), 24 November 1999.

Imai, Ryukichi, 'Global governance: some reflections', IIPS Policy Paper 191E, Tokyo: Institute for International Policy Studies, December 1997.

Institute for Human Ecology, *China Environment and Development Report*, 1(1), 1997.

Israel, Arturo, *Institutional Development: Incentives to Performance*, Baltimore, MD: Johns Hopkins University Press, 1988.

Jacobs, Michael, 'Sustainability and "the market": a typology of environmental economics', in Robyn Eckersley (ed.), *Markets, the State and the Environment: Towards Integration*, South Melbourne: Macmillan Education Australia, 1995, pp. 46–70.

Jahiel, Abigail R., 'Policy implementation through organizational learning: the case of water pollution management in China's reforming socialist system', PhD dissertation, University of Michigan, 1994.

——, 'The contradictory impact of reform on environmental protection in China', *China Quarterly*, 149(March), 1997, pp. 81–103.

——, 'The organization of environmental protection in China', *China Quarterly*, 156(December), 1998, pp. 757–87.

Jänicke, Martin, *Preventative Environmental Policy as Ecological Modernisation and Structural Policy*, Berlin: Berlin Science Center, 1985.

——, 'Conditions for environmental policy success: an international comparison', *The Environmentalist*, 12(1), 1992, pp. 47–58.

Jänicke, Martin and Hans Weidner (eds), *National Environmental Policies: A Comparative Study of Capacity-Building*, Berlin: Springer, 1997.

Japan Aid Study Group on Environment, *Japan International Cooperation Agency Sectoral Study for Development Assistance*, Tokyo: JICA, December 1988.

Japan–China Expert Committee, 'Nitchû kankyô kaihatsu moderu toshi kôsô ni kansuru teigen' (Policy recommendations for the Japan–China environmental model city project), unpublished draft report, Tokyo: JICA, 1998.

Japan, Ministry of Foreign Affairs, *Japan's Official Development Assistance Annual Report 1995*, Tokyo: Association for the Promotion of International Cooperation, 1996.

——, *Japan's Official Development Assistance Annual Report 1996*, Tokyo: Association for the Promotion of International Cooperation, 1997.

——, *Japan's Official Development Assistance Annual Report 1997*, Tokyo: Association for the Promotion of International Cooperation, 1998.

——, *Japan's Official Development Assistance Annual Report 1998*, Tokyo: Association for the Promotion of International Cooperation, 1999.

——, *Japan's Official Development Assistance Annual Report 1999*, Tokyo: Association for the Promotion of International Cooperation, 2000.

——, 'Economic cooperation program for China', www.mofa.go.jp/policy/oda/region/ e_asia/china-2.html (accessed 3 March 2002).

JBIC (Japan Bank for International Cooperation), *JBIC Annual Report*, Tokyo: JBIC, 1999.

——, *JBIC Annual Report*, Tokyo: JBIC, 2000.

——, *JBIC Annual Report*, Tokyo: JBIC, 2001.

——, *JBIC Annual Report*, Tokyo: JBIC, 2002.

——, *JBIC Annual Report*, Tokyo: JBIC, 2003.

——, 'Supporting environmental conservation and human resource development in China – FY2003 ODA loan package for China', www.jbic.go.jp/autocontents/english/news/ 2004/000025/index.htm (accessed 5 October 2004 and 13 January 2005).

——, 'List of anti-global warming projects (FY1998-FY2003)', www.jbic.go.jp/english/ environ/support/overseas/warming.php (accessed 21 March 2005).

JICA (Japan International Cooperation Agency), 'Report of the second country study for Japan's official development assistance to the People's Republic of China – findings and recommendations', unpublished, Tokyo: JICA, 1999.

Jolly, Richard, Giovanni Cornia and Frances Steward (eds), *Adjustment with a Human Face*, Oxford: Clarendon Press, 1989.

Kanda, Hiroshi, 'A big lie: Japan's ODA and environmental policy', *AMPO: Japan–Asia Quarterly Review*, 23(3), 1992, pp. 42–5.

Kapp, William, *Environmental Policies and Development Planning in Contemporary China and Other Essays*, Paris: Mouton, 1974.

Kapur, Devesh, John Lewis and Richard Webb, 'Introduction', in Devesh Kapur, John Lewis and Richard Webb (eds), *The World Bank: Its First Half Century*, volume 1, Washington, DC: Brookings Institution, 1997, pp. 1–56.

—— (eds), *The World Bank: Its First Half Century*, volumes 1 and 2, Washington, DC: Brookings Institution, 1997.

Katô Kazuo, *The Use of Market-Based Instruments in Japanese Environmental Policy*, Tokyo: Japan Environment Agency, 1993.

Korten, David, *Getting to the 21st Century: Voluntary Action and the Global Agenda*, West Hartford, CT: Kumarian Press, 1990.

Kothari, Rajni, *Rethinking Development: In Search of Humane Alternatives*, New York: New Horizons Press, 1989.

Kusano Atsushi, 'Japan's ODA in the 21st century', *Asia Pacific Review*, 7(1), 2000, pp. 38–55.

Lardy, Nicholas, *Agricultural Prices in China's Modern Economic Development*, Cambridge: Cambridge University Press, 1983.

Leading Group Office for Implementing China's Agenda 21, *1998 Wuhan chengshi ke chixu fazhan huanjing wenti zixun dahui* (Sustainable Development Environmental Issues Consultation Conference in Wuhan 28–30 April 1998), Wuhan: Wuhan Planning Committee, 1998.

Leading Office for Implementing China's Agenda 21, *Wuhan Environment Profile*, Wuhan: Agenda 21 Office, 1998.

Lim, Louise, 'China warns of water pollution', BBC news report, Beijing, 23 March 2005, http://news.bbc.co.uk/2/hi/asia-pacific/4374383.stm (accessed 27 March 2005).

Lipschutz, Ronnie and Ken Conca (eds), *The State and Social Power in Global Environmental Politics*, New York: Columbia University Press, 1993.

Liu Changming, 'Environmental issues and the south–north water transfer scheme', *China Quarterly*, 156(December), 1998, pp. 899–910.

Liuzhou Environmental Protection Bureau, 'The situation of ambient air pollution in Liuzhou', unpublished, Liuzhou: EPB, 1998.

Lo, Wing-Hung Carlos and Shui-Yan Tang, 'Institutional contexts of environmental management: water pollution control in Guangzhou, China', *Public Administration and Development*, 14, 1994, pp. 53–64.

Lorenz, Andreas, 'China's environmental suicide: a government minister speaks', interview with Pan Yue, 5 April 2005, www.opendemocracy.net/debates/article-6-129-2407.jsp (accessed 8 April 2005).

Lotspeich, Richard and Aimin Chen, 'Environmental protection in the People's Republic of China', *Journal of Contemporary China*, 6(14), 1997, pp. 33–59.

Luk, Shiu-Hung and Joseph Whitney (eds), *Megaproject: A Case Study of China's Three Gorges Dam Project*, Armonk, NY: M.E. Sharpe, 1993.

Ma, Xiaoying and Leonard Ortolano, *Environmental Regulation in China: Institutions, Enforcement, and Compliance*, Lanham, MD: Rowman and Littlefield, 2000.

McCormack, Gavan, *The Emptiness of Japanese Affluence*, Armonk, NY: M.E. Sharpe, 1996.

McElroy, Michael B., 'Industrial growth, air pollution, and environmental damage: complex challenges for China', in Michael B. McElroy, Chris P. Nielsen and Peter Lydon (eds), *Energizing China: Reconciling Environmental Protection and Economic Growth*, Cambridge, MA: Harvard University Press, 1998, pp. 241–65.

McElroy, Michael B., Chris P. Nielsen and Peter Lydon (eds), *Energizing China: Reconciling Environmental Protection and Economic Growth*, Cambridge, MA: Harvard University Press, 1998.

McHale, John and Magda McHale, *Basic Human Needs*, New Brunswick, NJ: Transaction Books for UNEP, 1978.

MacNeill, Jim, Pieter Winsemius and Taizo Yakushiji, *Beyond Interdependence: The Meshing of the World's Economy and the Earth's Ecology*, New York: Oxford University Press, 1991.

Maddock, Rowland T., 'Japan and global environmental leadership', *Journal of Northeast Asian Studies*, 13(4), 1994, pp. 37–48.

Marglin, Appel and Stephen Marglin (eds), *Dominating Knowledge: Development, Culture and Resistance*, Oxford: Clarendon Press, 1990.

Margulis, Sergio and Tonje Vetleseter, 'Environmental capacity building – a portfolio review', *Environment Matters at the World Bank: Annual Review*, Washington, DC: World Bank, Fall 1998.

Mason, Edward and Robert Asher, *The World Bank since Bretton Woods*, Washington, DC: Brookings Institution, 1973.

Matsuura Shigenori, 'China's air pollution and Japan's response to it', *International Environmental Affairs*, 7(3), 1995, pp. 235–48.

Miller, Lyman, 'China's leadership transition: the first stage', *China's Leadership Monitor*, 5, 2003.

Mol, Arthur, 'Ecological modernisation and institutional reflexivity: environmental reform in the late modern age', *Environmental Politics*, 5(2), 1996, pp. 302–23.

Moore, Mick, 'Toward a useful consensus', *IDS Bulletin*, 29(2), 1998, pp. 39–48.

Moran, Alan, Andrew Chisholm and Michael Porter (eds), *Markets, Resources and the Environment*, North Sydney: Allen and Unwin, 1991.

Morton, Katherine, 'Transnational advocacy at the grassroots in China: potential benefits and risks,' *China Information*, 20(1 and 2), 2006.

Mosley, Paul, Jane Harrigan and John Toye, *Aid and Power: The World Bank and Policy-Based Lending*, volumes 1 and 2, London: Routledge, 1991.

Myers, Norman, 'China's approach to environmental conservation', *Environmental Affairs*, 5(33), 1976, pp. 33–63.

Naidi, Zhang, 'Automobile exhaust pollution and its control in Wuhan', in Leading Office for Implementing China's Agenda 21, *Wuhan Environment Profile*, Wuhan: Agenda 21 Office, 1998.

National Geographic, *Atlas of the World*, 5th edn, Washington, DC: National Geographic Society, 1981.

National Resources Defense Council, 'Second analysis confirms greenhouse gas reductions in China', www.nrdc.org/globalwarming/achinagg.asp (accessed 5 April 2002).

Naughton, Barry, 'China: domestic restructuring and a new role in Asia', in T.J. Pempel (ed.), *The Politics of the Asian Economic Crisis*, Ithaca, NY: Cornell University Press, 1999, pp. 203–23.

Needham, Joseph, *Science and Civilisation in China*, Cambridge: Cambridge University Press, 1965.

——, *Science in Traditional China: A Comparative Perspective*, Cambridge, MA: Harvard University Press, 1981.

Nelson, Joan M. and Stephanie J. Eglinton, *Global Goals, Contentious Means: Issues of Multiple Aid Conditionality*, Washington, DC: Overseas Development Council, 1993.

Newcombe, Ken, Juergen Blaser and Kerstin Canby, 'The World Bank and forests', in *Environment Matters at the World Bank: Annual Review*, Washington, DC: World Bank, Fall 1998, pp. 56–9.

Nishikawa, Jun, 'Reform of foreign aid', *Nikkei Weekly*, 12 May 1997.

Nolla, Eduardo, *De la democratie en Amerique*, Paris: J. Vrin, 1990.

OECD (Organisation for Economic Cooperation and Development), *Environmental Priorities for China's Sustainable Development*, Paris: OECD, 10 December 2001.

OECD, DAC (Development Assistance Committee), *1994 Environmental Policy Review*, DAC Report, Paris: OECD, 1995.

OECD, DAC, *Development Cooperation: efforts and policies of the members of the Development Assistance Committee*, DAC Report, Paris: OECD, 1996.

OECF (Overseas Economic Cooperation Fund), Special Assistance for Project Formation Study, 'People's Republic of China environmental improvement project – interim report of China environmental improvement project', unpublished, Tokyo: OECF, March 1995.

——, Quarterly Report, *Outline of 40 Planned Projects for Fourth Batch of ODA Loans to China*, Tokyo: OECF, August 1997.

——, Newsletter, *Toward Improvement of the Environment and Alleviation of Poverty*, Tokyo: OECF, March 1999.

Oi, Jean C., 'Fiscal reform and the economic foundations of local state corporatism in China', *World Politics*, 45(1), 1992, pp. 99–126.

Olson, Mancur, *The Logic of Collective Action*, Cambridge, MA: Harvard University Press, 1965.

Ophuls, William, 'Leviathan or oblivion', in Herman Daly (ed.), *Toward a Steady-State Economy*, San Francisco, CA: Freeman, 1973, pp. 215–30.

Orleans, L. and Richard Suttmeier, 'The Mao ethic and environmental quality', *Science*, 170, 1970, pp. 1173–6.

Panayotou, Theodore, 'Economic incentives for environmental management in developing countries', in OECD, *Economic Instruments for Environmental Management in Developing Countries*, Paris: OECD, 1992.

——, *Instruments of Change: Motivating and Financing Sustainable Development*, London: Earthscan, 1998.

Pearce, David, Anil Markandya and Edward Barbier, *Blueprint for a Green Economy*, London: Earthscan, 1989.

Pharr, Susan and Ming Wan, 'Yen for the earth: Japan's pro-active China environmental policy', in Michael B. McElroy, Chris P. Nielsen and Peter Lydon (eds), *Energizing China: Reconciling Environmental Protection and Economic Growth*, Cambridge, MA: Harvard University Press, 1998, pp. 601–38.

Piddington, Kenneth, 'The role of the World Bank', in Andrew Hurrell and Benedict Kingsbury (eds), *The International Politics of the Environment: Actors, Interests, and Institutions*, Oxford: Clarendon Press, 1992, pp. 212–27.

Pigou, Arthur, *The Economics of Welfare*, London: Macmillan, 1932, first published 1920.

Potter, David, 'Assessing Japan's environmental aid policy', *Pacific Affairs*, 67(2), 1994, pp. 200–15.

Pye, Lucian, *Aspects of Political Development: An Analytic Study*, Boston, MA: Little Brown, 1966.

Qu Geping, 'China's industrial pollution survey', *China Reconstructs*, 37(8), 1988, pp. 16–18.

——, 'China's environmental policy and world environmental problems', *International Environmental Affairs*, 2(2), 1990, pp. 103–8.

Qu Geping and Li Jinchang, *Population and Environment in China*, Boulder, CO: Lynne Rienner, 1994.

Redclift, Michael, 'Sustainable development and popular participation: a framework for analysis', in Dharam Ghai and Jessica Vivian (eds), *Grassroots Environmental Action: People's Participation in Sustainable Development*, New York: Routledge, 1992, pp. 23–49.

Reddy, Amulya, Robert Williams and Thomas Johansson, *Energy After Rio: Prospects and Challenges*, New York: UNDP, 1997.

Reed, David, *Structural Adjustment and the Environment*, Boulder, CO: Westview Press, 1992.

Rich, Bruce M., *Mortgaging the Earth: The World Bank, Environmental Impoverishment, and the Crisis of Development*, Boston, MA: Beacon Press, 1994.

Richardson, Stanley Dennis, *Forests and Forestry in China: Changing Patterns of Resource Development*, Washington, DC: Island Press, 1990.

Rivkin, Arnold (ed.), *Nations by Design: Institution Building in Africa*, Garden City, NY: Doubleday, 1968.

Rix, Alan, *Japan's Foreign Aid Challenge: Policy Reform and Aid Leadership*, London and New York: Routledge, 1993.

Rozelle, Scott, Xiaoyang Ma and Leonard Ortolano, 'Industrial wastewater control in Chinese cities: determinants of success in environmental policy', *Natural Resources Modeling*, 7(4), 1993, pp. 353–78.

Ruttan, Lore, 'Closing the commons: cooperation for gain or restraint?', *Human Ecology*, 26(1), 1998, pp. 43–66.

Sagoff, Mark, *The Economy of the Earth: Philosophy, Law, and the Environment*, Cambridge: Cambridge University Press, 1988.

Saich, Tony, 'Negotiating the state: the development of social organizations in China', *China Quarterly*, 161(March), 2000, pp. 124–41.

Saito, Masato, *Country Study Group for Development Assistance to the People's Republic of China: Basic Strategy for Development Assistance*, Tokyo: JICA, December 1991.

Sandbrook, Richard, 'UNGASS has run out of steam', *International Affairs*, 73(4), 1997, pp. 641–54.

Schreurs, Miranda and Y. Peng, 'The Earth Summit and Japan's initiative in environmental diplomacy', *Futures*, 25(4), 1993, pp. 379–91.

Schumacher, Freidrich Ernst, *Small is Beautiful: A Study of Economics as if People Mattered*, London: Blond and Briggs, 1973.

Schwartzman, Stephan, *Bankrolling Disasters: International Banks and the Global Environment*, San Francisco, CA: Sierra Club, 1986.

Searle, Graham, *Major World Bank Projects: Their Impact on People, Society and the Environment*, Canelford, Cornwall: Wadebridge Ecological Centre, 1987.

Sen, Amartya, 'Public action and the quality of life in developing countries', *Oxford Bulletin of Economics and Statistics*, 43(4), 1981, pp. 287–319.

——, 'The ends and means of development', in Amartya Sen (ed.), *Development as Freedom*, Oxford: Oxford University Press, 1999, pp. 35–53.

SEPA (State Environmental Protection Administration), 'Riyuan daikuan huanbao xiangmu jishu yu xinxi jiaoliu zhinan' (Japanese yen loan environmental protection projects: technical and information guide), unpublished, Beijing: SEPA, 1999.

Serageldin, Ismail, *Water Supply, Sanitation, and Environmental Sustainability: The Financing Challenge*, World Bank Directions in Development Series, Washington, DC: World Bank, 1994.

Shafritz, Jay M., *Dictionary of Public Administration*, New York: Oxford University Press, 1986.

Shapiro, Judith, *Mao's War Against Nature: Politics and the Environment in Revolutionary China*, Cambridge: Cambridge University Press, 2001.

Shen, Xiao-ming, John F. Rosen, Di Guo Wu and Sheng-mei, 'Childhood lead poisoning in China', *The Science of the Total Environment*, 181(2), 1996, pp. 101–9.

Shenyang chengshi ke chixu fazhan xiangmu: ke chixu fazhan de jintian he mingtian (Shenyang Sustainable City Development Project: sustainable development today and tomorrow), Shenyang: Huanjing baohu ju, 1997.

Shenyang Environmental Protection Bureau, 'The profile of Shenyang environmental protection', unpublished, Shenyang: Shenyang EPB, May 1999.

Shenyang Planning Committee, 'Shenyang shi liyong riyuan daikuan shishi daqi wuran zhili xiangmu qingkuang jieshao' (Introduction to Shenyang City implementation of the Japanese yen loan project to control air pollution), unpublished, Shenyang: Shenyang Green Project Office, 1999.

Shenyang Sustainable City Project, *Shenyang Environmental Profile*, Shenyang: Sustainable Shenyang Project Office, April 1998.

——, *Chengshi zixun dahui. Daibiao shouce* (City consultation conference: participants manual), Shenyang: Sustainable Shenyang Project Office, 5–7 May 1998.

Shepherd, Andrew, 'Participatory environmental management: contradiction of process, project and bureaucracy in the Himalayan foothills', *Public Administration and Development*, 15(4), 1995, pp. 465–79.

Shihata, Ibrahim, *The World Bank in a Changing World*, Dordrecht: Martinus Nijhoff Publishers, 1991.

Shirk, Susan, *The Logic of Economic Reform in China*, Berkeley, CA: University of California Press, 1993.

Shiroyama, Hideaki, 'Clean and efficient coal use in China and the political economy of international aid', *Social Science Japan*, 16, August 1999, pp. 15–19.

Sinkule, Barbara and Leonard Ortolano, *Implementing Environmental Policy in China*, Westport, CT: Praeger, 1995.

Sitarz, Paul and Daniel Sitarz (eds), *Agenda 21: The Earth Summit Strategy to Save Our Planet*, Boulder, CO: Earth Press, 1994.

Smil, Vaclav, *The Bad Earth: Environmental Degradation in China*, Armonk, NY: M.E. Sharpe, 1984.

——, *China's Environmental Crisis: An Inquiry into the Limits of National Development*, Armonk, NY: M.E. Sharpe, 1993.

——, 'Environmental problems in China: estimates of economic costs', Special Report No. 5, Honolulu: East West Center, April 1996.

——, *China's Past, China's Future: Energy, Food, Environment*, New York: Routledge, 2003.

Spaargaren, Gert and Arthur Mol, 'Sociology, environment, and modernity: ecological modernization as a theory of social change', *Society and Natural Resources*, 5(4), 1992, pp. 323–44.

State Council, *China's Agenda 21: White Paper on China's Population, Environment, and Development in the 21st Century*, Beijing: Zhongguo huanjing kexue chubanshe, 1994.

——, White Paper, *The Development-Oriented Poverty Reduction Program*, Beijing: State Council, 15 October 2001.

Stern, Nicholas and Francisco Ferreira, 'The World Bank as "intellectual actor"', in Devesh Kapur, John Lewis and Richard Webb (eds), *The World Bank: Its First Half Century*, volume 2, Washington, DC: Brookings Institution, 1997, pp. 523–610.

Story, Greg, 'Japan's official development assistance to China: a survey, Pacific Economic Papers No. 150, Canberra: Australia Japan Research Centre, Australian National University, 1987.

Streeten, Paul, Shahid Javed Burki, Mahbub ul-Haq, Norman Hicks and Frances Stewart, *First Things First*, Oxford: Oxford University Press, 1981.

Study Group for Global Environment and Economics, *Pollution in Japan – Our Tragic Experience: Case Studies of Pollution-Related Damage at Yokkaichi, Minamata, and the Jinzu River*, Tokyo: Environment Agency, July 1991.

Tang, Shui-Yan, Carlos Wing-Hung Lo, Kai-Chee Lo Cheung and Jack Man-Keung, 'Institutional constraints on environmental management in urban China: environmental impact assessment in Guangzhou and Shanghai', *China Quarterly*, 152(December), 1997, pp. 863–74.

Taylor, Jonathan, 'Japan's global environmentalism: rhetoric and reality', *Political Geography*, 18(5), 1999, pp. 535–62.

Tietenberg, Tom H., *Emissions Trading: An Exercise in Reforming Pollution Policy*, Washington, DC: Resources for the Future, 1985.

Tietenberg, Tom H., 'Economic instruments for environmental regulation', *Oxford Review of Economic Policy*, 6(1), 1990, pp. 17–33.

Tietenberg, Tom H. and David Wheeler, 'Empowering the community: information strategies for pollution control', paper presented at the Frontiers of Environmental Economics Conference, Airlie House, Virginia, 23–25 October 1998.

Timoshenko, Alexander and Mark Berman, 'The United Nations Environment Programme and the United Nations Development Programme', in Jacob Werksman (ed.), *Greening International Institutions*, London: Earthscan, 1996, pp. 38–54.

Tolba, Mostafa, *Development without Destruction: Evolving Environmental Perceptions*, Dublin: Tycooly International, 1982.

Topping, Audrey R., 'Ecological roulette: damming the Yangtze', *Foreign Affairs*, 74(5), 1995, pp. 132–46.

Turner Jennifer, 'Cultivating environmental NGO–business partnerships', *China Business Review*, November–December, 2003.

Ui, Jun, 'Minamata disease and Japan's development', *AMPO: Japan–Asia Quarterly Review*, 27(3), 1997, pp. 18–25.

UNDP (United Nations Development Programme), *Protecting the Environment*, New York: UNDP, 1988.

——, *UNDP Initiative for Sustainable Energy*, New York: UNDP, June 1996.

——, *Annual Report 1996/1997, Ending Poverty and Building Peace through Sustainable Human Development*, New York: UNDP, October 1997.

——, *Empowering People: A Guide to Participation*, New York: UNDP, 1997.

——, *UNDP Today: Introducing the Organisation*, New York: UNDP, September 1998.

——, The Fifth Year: Learning and Growing, Annual Report Capacity 21, New York: UNDP, 1998.

——, 'GEF portfolio for China', unpublished, New York: UNDP, March 1999.

——, *Annual Report, Partnerships to Fight Poverty*, New York: UNDP, 2001.

——, *Capacity 21 Annual Report*, New York: UNDP, 2001.

——, *Annual Report 2003, A World of Development Experience*, New York: UNDP, 2003.

UNDP China, 'Capacity building for widespread adoption of clean production for air pollution control in Benxi', Project No. CPR/96/307/B/01/99, unpublished, Beijing: UNDP, 1996.

——, 'Capacity development for acid rain and SO_2 pollution control in Guiyang', Project No. CPR/96/304/A/01/99, unpublished, Beijing: UNDP, 1996.

——, *China Environment and Sustainable Development Resource Book II: A Compendium of Donor Activities*, Beijing: UNDP, 1996.

——, 'Managing sustainable development in Shenyang', Project No. CPR/96/321/A/01/99, unpublished, Beijing: UNDP, 1996.

——, *Environment and Sustainable Energy Development: A Strategy and Action Plan for UNDP in China*, Beijing: UNDP, November 1997.

——, 'Summary report of the evaluation mission: China urban pollution control program', Beijing: UNDP, May 2001.

UNDP and UNDESA (United Nations Department of Economic and Social Affairs), *The Challenge of Sustainability*, New York: UNDP, 2000.

UNDP and UNDESA, *World Energy Assessment: Energy*, New York: UNDP, 2000.

UNDP and UNPF (United Nations Population Fund), *Results-Oriented Annual Report (ROAR)*, New York: UNDP, 2000.

UNDP and UNPF, *Second Country Cooperative Framework for China (2001–2005)*, New York: UNDP, September 2001.

UNDP and UNPF, *The Results-Oriented Annual Report 2001*, New York: UNDP, 2001.

Unger, Jonathan, ' "Bridges": private business, the Chinese government and the rise of new associations', *China Quarterly*, 147(September), 1996, pp. 795–819.

——, *The Transformation of Rural China*, Armonk, NY: M.E. Sharpe, 2002.

Uphoff, Norman, *Local Institutional Development*, West Hartford, CT: Kumarian Press, 1986.

Vermeer, Eduard B., 'Industrial pollution in China and remedial polices', *China Quarterly*, 156(December), 1998, pp. 952–85.

——, 'Management of environmental pollution in China: problems and abatement measures', *China Information*, 5(1), 1990, pp. 32–65.

Wade, Robert, 'Greening the Bank: the struggle over the environment, 1970–1995', in Devesh Kapur, John Lewis and Richard Webb (eds), *The World Bank: Its First Half Century*, volume 2, Washington, DC: Brookings Institution, 1997, pp. 611–734.

——, 'The market for public office: why the Indian state is not better at development', *World Development*, 13(4), 1985, pp. 467–97.

Wang, Hanchen and Liu Bingjiang, 'Policymaking for environmental protection in China', in Michael B. McElroy, Chris P. Nielsen and Peter Lydon (eds), *Energizing China: Reconciling Environmental Protection and Economic Growth*, Cambridge, MA: Harvard University Press, 1998, pp. 371–403.

Wang, Xiaolu, 'Economic growth over the past twenty years', in Ross Garnaut and Ligang Song (eds), *China: Twenty Years of Economic Reform*, Canberra: Asia Pacific Press, 1999, pp. 27–50.

Wapenhams, Willi, *Report of the Portfolio Management Task Force*, Washington, DC: World Bank, 1992.

Weale, Albert, *The New Politics of Pollution*, Manchester: Manchester University Press, 1992.

Wenger, Robert, Wang Huadong and Ma Xiaoying, 'Environmental impact assessment in the People's Republic of China', *Environmental Management*, 14(4), 1990, pp. 429–39.

White, Gordon, *Riding the Tiger: The Politics of Economic Reform in Post-Mao China*, Basingstoke: Macmillan, 1993.

White, Gordon, Jude Howell and Shang Xiaoyuan, *In Search of Civil Society: Market Reform and Social Change in Contemporary China*, Oxford: Clarendon Press, 1996.

Wilson, Elizabeth, 'Capacity for environmental action in Slovakia', *Journal of Environmental Planning and Management*, 42(4), 1999, pp. 581–98.

Wolfensohn, James, 'The other crisis', Address to the Board of Governors, Washington, DC: World Bank, 6 October 1998.

World Bank, *China: Efficiency and Environmental Impact of Coal*, Washington, DC: World Bank, 1991.

——, *World Bank Annual Report on the Environment*, Washington, DC: World Bank, 1991.

——, *World Bank Development Report: Development and the Environment*, Washington, DC: World Bank, 1992.

——, *World Bank Development Report: China: Involuntary Resettlement*, Washington, DC: World Bank, 1993.

——, *World Bank Development Report: China Urban Environmental Service Management*, Report No. 13073-CHA, Washington, DC: World Bank, 1994.

——, *Japan's Experience in Urban Environmental Management*, Washington, DC: Metropolitan Environment Improvement Program, 1995.

——, *China 2020: Development Challenges in the New Century*, Washington, DC: World Bank, 1997.

——, *Clear Water and Blue Skies: China's Environment in the Twenty-First Century*, Washington, DC: World Bank, 1997.

——, *Can the Environment Wait?*, Washington, DC: World Bank, 1997.

World Bank, *World Development Report 1997: The State in a Changing World*, Washington, DC: World Bank, 1997.

——, 'Environmental projects portfolio', in *Environment Matters at the World Bank: Annual Review*, Washington, DC: World Bank, Fall 1998.

——, *Managing Pollution Problems*, Washington, DC: World Bank, 1998.

——, *Transition Towards a Healthier Environment: Environmental Issues and Challenges in the Newly Industrialised States*, Washington, DC: World Bank, 1998.

——, *China, Air, Land, and Water: Environmental Priorities for a New Millennium*, Washington, DC: World Bank, 2001.

——, *Making Sustainable Commitments: An Environment Strategy for the World Bank*, Washington, DC: World Bank, 2001.

——, 'The Bank's environment portfolio', in *Environment Matters at the World Bank*, Washington, DC: World Bank, June 2001.

——, 'Memorandum of the president of IBRD and IFC to the executive directors in a Country Assistance Strategy of the World Bank Group for the People's Republic of China', Washington, DC: China Country Management Unit, East Asia and Pacific Region, 22 January 2003.

——, *Environment Matters at the World Bank: Annual Review*, Washington, DC: World Bank, July 2002–June 2003.

——, *World Bank Annual Report 2003*, Washington, DC: World Bank, 2003.

——, *Clean Development Mechanism in China: Taking a Proactive and Sustainable Approach*, Washington, DC: International Bank for Reconstruction and Development/ The World Bank, June 2004.

——, *Implementation Completion Report on a Loan in the Amount of US$110 Million to the People's Republic of China for a Liaoning Environment Project*, Report No. 28833-CHA, Washington, DC: Urban Development Sector Unit, East Asia and Pacific Region, 22 June 2004.

——, *Focus on Sustainability 2004*, Washington, DC: International Bank for Reconstruction and Development/World Bank, January 2005.

World Bank, China, *The World Bank Group in China, Facts and Figures*, Beijing: World Bank, 1999.

——, *The World Bank Group in China, Facts and Figures*, Beijing: World Bank, 2002.

——, *The World Bank Group in China, Facts and Figures*, Beijing: World Bank, 2003.

World Bank, Environmentally Sustainable Development, *The World Bank Participation Sourcebook*, Washington, DC: World Bank, 1996.

World Bank Staff Appraisal Report No. 12708-CHA, *China Liaoning Environment Project*, Washington, DC: Urban Development Sector Unit, East Asia and Pacific Regional Office, July 1994.

World Bank Staff Appraisal Report No. 15361-CHA, *China Yunnan Environment Project*, Washington, DC: Urban Development Sector Unit, East Asia and Pacific Regional Office, May 1996.

World Bank Staff Appraisal Report No. 16622-CHA, *China Guangxi Urban Environment Project*, Washington, DC: Urban Development Sector Unit, East Asia and Pacific Regional Office, May 1998.

World Commission on Environment and Development, *Our Common Future*, Oxford: Oxford University Press, 1987.

Wu, Fenshi, 'New partners or old brothers? GONGOs in transnational environmental advocacy in China', *China Environment Series*, 5, 2002, pp. 45–58.

Wu Jiachun, 'Yunnan battles to save the Dianchi Lake', *China Daily*, 29 October 1998.

Wu Ming, 'Disaster in the making? major problems found in the Three Gorges Dam resettlement', *China Rights Forum: The Journal of Human Rights in China*, Spring, 1998, pp. 4–9.

Xue Wei, 'Water resources and economic development in China', *Chinese Economic Studies*, 29(1), 1996.

Yasutomo, Dennis, *The Manner of Giving: Strategic Aid and Japanese Foreign Policy*, Lexington, MA: Lexington Books, 1986.

Yoda, Susumu, 'Japan's international contribution', keynote speech presented at the International Conference on Energy and Sustainable Development, Tsinghua University, Beijing, 16–17 July 1996.

Young, Nick, 'China always in the driving seat', *China Development Briefing*, 1(4), 1997.

——, ' "Ethnic diluting" row clouds last soft bank loan', *China Development Brief*, 2(3), 1999, pp. 32–3.

——, 'Bilateral douors trim budgets, advance governance and policy agendas', *China Development Brief*, 7(1), Spring 2004.

Yunnan Environmental Protection Bureau, *Yunnan sheng huanjing baohu jiuwu jihua he 2010 nian chang yuan guihua* (Yunnan Province Environmental Protection Ninth Five-Year Plan and Long Term Plan until 2010), Kunming: Yunnan EPB, 1996.

——, *Yunnan sheng shijie yinhang daikuan huanbao xiangmu jieshao* (Introduction to the Yunnan World Bank Environmental Loan Project), Kunming: Yunnan EPB, 1996.

Yunnan Institute of Environmental Sciences, *Environmental Assessment Report*, Kunming: Yunnan EPB, February 1996.

Zhang Weijong, Ilan Vertinsky, Terry Ursacki and Peter Nemetz, 'Can China be a clean tiger?: Growth strategies and environmental realities', *Pacific Affairs*, 72(1), 1999, pp. 23–37.

Zhang, Z.X. and Henk Folmer, 'The Chinese energy system: implications for future carbon dioxide emissions in China', *Journal of Energy and Development*, 21(1), 1996, pp. 1–44.

Zhongguo huanjing kexue (China Environmental Science), October 1997.

Zhongguo huanjing nianjian 1995 (China Environmental Yearbook 1995), Beijing: Zhongguo huanjing chubanshe, 1995.

Zhongguo huanjing nianjian 1996 (China Environmental Yearbook 1996), Beijing: Zhongguo huanjing chubanshe, 1996.

Zhongguo huanjing nianjian 1997 (China Environmental Yearbook 1997), Beijing: Zhongguo huanjing chubanshe, 1997.

Zhongguo huanjing nianjian 2000 (China Environmental Yearbook 2000), Beijing: Zhongguo huanjing chubanshe, 2000.

Zhongguo tongji nianjian 1988 (China Statistical Yearbook 1988), Beijing: Zhongguo tongji chubanshe, 1988.

Zhongguo tongji nianjian 1995 (China Statistical Yearbook 1995), Beijing: Zhongguo tongji chubanshe, 1995.

Zhongguo tongji nianjian 1996 (China Statistical Yearbook 1996), Beijing: Zhongguo tongji chubanshe, 1996.

Zhongguo tongji nianjian 1997 (China Statistical Yearbook 1997), Beijing: Zhongguo tongji chubanshe, 1997.

Zhongguo tongji nianjian 2000 (China Statistical Yearbook 2000), Beijing: Zhongguo tongji chubanshe, 2000.

Ziegler, Charles (ed.), *Environmental Policy in the USSR*, Amherst, MA: University of Massachusetts, 1987.

Index

For Product Safety Concerns and Information please contact our EU
representative GPSR@taylorandfrancis.com
Taylor & Francis Verlag GmbH, Kaufingerstraße 24, 80331 München, Germany